Guide to
Information Sources in
the Botanical Sciences

Reference Sources in Science and Technology Series

Guide to the Literature of Pharmacy and the Pharmaceutical Sciences. By Theodora Andrews.

Guide to the Petroleum Reference Literature. By Barbara C. Pearson and Katherine B. Ellwood. Foreword by Arthur R. Green.

Guide to Information Sources in the Botanical Sciences. By Elisabeth B. Davis.

Guide to
Information Sources in
the Botanical Sciences

Elisabeth B. Davis

1987

Libraries Unlimited, Inc.   Littleton, Colorado

LIBRARIES UNLIMITED, INC.
P.O. Box 263
Littleton, Colorado 80160-0263

Library of Congress Cataloging-in-Publication Data

Davis, Elisabeth B., 1932-
    Guide to information sources in the botanical
sciences.

    (Reference sources in science and technology series)
    Includes bibliographies and index.
    1. Botany--Bibliography. 2. Reference books--
Botany--Bibliography. 3. Botany--Information services--
Directories. 4. Information storage and retrieval
systems--Botany--Directories. I. Title. II. Series.
Z5351.D38 1987 [QK45.2]   016.58   86-27574
ISBN 0-87287-439-7

Contents

Part I
Bibliographic Control

Part II
Ready-Reference Sources

Part III
Additional Sources of Information

Preface

The purpose of this guide is to provide a useful survey of information sources for students, librarians, avocational and professional botanists. The treatment is comprehensive in scope, ranging from materials selected for the informed layperson to those for the specialist, with selections appropriate for public, college, university, and professional libraries. Botany is broadly construed to encompass the plant kingdom in all its divisions and in all its aspects, except those of agriculture, horticulture, and gardening. Emphasis is on English-language materials, although foreign-language materials are included whenever necessary to provide complete coverage, especially for retrospective sources. There are no chronological restrictions; out-of-print (o.p.) works are included when appropriate, but the majority of the annotations are for currently available reference works. Teaching of botany is not included, nor are instructions for conducting library research, although some reference sources for library research techniques are annotated with guides to the literature in chapter 1. The major focus is on printed materials and computerized databases with little or no attention given to audiovisual materials, film catalogs, and the like used for instructional purposes.

All materials annotated in this guide are recommended; selection criteria did not allow the inclusion of reference works of ephemeral character or dubious utility. Reference sources describing biotechnology, patents, and translations are included because of their current importance and highly visible nature. Throughout the annotations, references to book reviews (noted "R") have been provided whenever possible to supplement evaluations. A list of publications cited follows this preface.

Prices, although quickly outdated, have been included whenever possible to indicate the price range one may expect to find in botanical information sources. Prices have been taken from 1985/1986 editions of standard bibliographic acquisition tools, or from the most recent price list circulated by the publisher. Materials with no prices given were not found in standard bibliographic sources and are assumed to be out-of-print. For some foreign publications, prices are given in U.S. dollars if they are listed that way in standard sources.

Acknowledgments are due to the University of Illinois Library Research and Publication Committee for financial aid, and to Dawn M. Jenkins, Vera Luner, Krzysztof Szymborski, and Jane E. Vinton for bibliographic assistance.

Publications Cited

Form of Citation	Publication Title
AAAS	AAAS: Science Books and Films
ALIN	Agricultural Libraries Information Notes
AMN	American Midland Naturalist
ARBA	American Reference Books Annual
BL	Booklist
Choice	Not abbreviated
CRL	College and Research Libraries
Huntia	Not abbreviated
IJSB	International Journal of Systematic Bacteriology
KewMag	Kew Magazine
LJ	Library Journal
NatHist	Natural History
Nature	Not abbreviated
Phyto	Phytologia
QRB	Quarterly Review of Biology
RQ	Reference Quarterly
Science	Not abbreviated
STBN	Scitech Book News
Winchell	Winchell, C. M. Guide to Reference Books. 8th ed. 1967.

Introduction to
the Botanical Literature

Botany, the study of plants, has been of great importance and a major influence on humankind ever since the first stirrings of intelligent life. The botanical sciences encompass areas from the applied to the theoretical, the descriptive to the functional, the ancient to the modern, the tiniest alga to the giant redwood. In the early centuries, human life was sustained by plants, which were used for food, drink, shelter, and weapons, and indirectly, for clothing. The domestication of plants and the origins of agriculture marked the beginnings of civilization. In *Basic Botany* (Cronquist, 1981), plants are defined as organisms that make their own food from raw materials of carbon dioxide and water and use light as an energy source and chlorophyll as an enzyme. This process, photosynthesis, converts energy to food and produces oxygen as a byproduct; because plants are the sole producers of food and oxygen, they are indispensable for life on Earth.

Classification of plant life has been important ever since ancient philosophers began organizing observations and facts into a knowledge framework. Plants almost always have been recognized as a distinct entity and quite different from animals or minerals. Commonly, life has been divided, for classification purposes, into two kingdoms, the Plant Kingdom and the Animal Kingdom. Various refinements have evolved until at the present time the most popular designations include either four or five kingdoms, depending on the criteria used. The classification scheme used in this book is based on four kingdoms: Virus, Monera, Plantae, and Animalia (Parker, 1982). Based on this classification, botany falls into Superkingdom Eukaryotae, Kingdom Plantae, and does not include virus or bacteria. Plantae is further divided into two

1

subkingdoms: the Thallobionta (eukaryotic algae and fungi) and Embryobionta (land plants).

The scope of the plant sciences is exceedingly broad. The four major botanical subdivisions of morphology, physiology, ecology, and systematics may be further broken down into plant anatomy, cytology, histology, development, biochemistry, biophysics, bacteriology, mycology, bryology, paleobotany, pathology, pharmacognozy, and other more esoteric specialities. The range of botanical impact covers the arts of weaving, painting, and decoration; scientific disciplines, including medicine, pharmacology, and organic chemistry; and applied work in agriculture, horticulture, and forestry. Important tools used in the study of botany include microscopy, staining techniques, gas and paper chromatography, electrophoresis, spectroscopy, other methods for biochemical and biophysical analysis, field work, carbon 14 dating, computer techniques, and organized research centers such as the herbarium, the botanical garden, and the library.

Although the Greek and Roman philosophers laid the early foundations for the science of botany, its economic and medicinal usefulness was established even earlier by the Assyrians, Chinese, Egyptians, and Hindus, who had knowledge of plant cultivation, medicinal and hallucinatory properties of plants, and production methods for food, beverages, and spices. Aristotle was one of the first Western philosophers who wrote about plants, but it was Theophrastus, Dioscorides, and Pliny the Elder who were the main botanical writers of antiquity. Theophrastus is generally given credit for being the founder of botany; Dioscorides wrote one of the earliest herbals describing and illustrating medicinal plants; Pliny the Elder produced the 37-volume *Historia Naturalis*, including 16 volumes on botany. Various Roman philosophers wrote on botanical subjects, too, but their emphasis was on farming, not scientific inquiry.

During the Dark Ages, scientific observation of all kinds declined, and it was not until the fifteenth and sixteenth centuries that the study of botany again flourished with the production of handbooks and catalogs by such naturalists as Otto Brunfels, Hieronymus Bock, Leonhard Fuchs, William Turner, and John Gerard, to name a few of the early herbalists. The seventeenth and eighteenth centuries were the ages of collectors and classifiers, and descriptive botany thrived in this atmosphere. During the eighteenth century, botany was established as a pure science, with inventions like the microscope, and experimenters and classifiers like Gaspard Bauhin, Robert Hooke, Nehemiah Grew, Marcello Malpighi, Stephen Hales, Joseph Priestley, Jan Ingen-Housz, John Ray, and Carolus Linnaeus making great contributions to plant anatomy, plant physiology, plant chemistry, and taxonomy.

Carolus Linnaeus marked a watershed in the history of botany. He was a teacher, a writer, a collector, a proselytizer with hundreds of correspondents and disciples, and an inventor of a system of binomial nomenclature still used today by modern plant taxonomists. His *Species Plantarum* (1753), together with his *Genera Plantarum* (5th ed., 1754), have generally been accepted by international agreement among botanists as the starting point for the nomenclature of plants in general. Linnaeus was not an experimentalist; in his view, the foundations of botany were twofold: classification and nomenclature, and he exemplified this spirit with enthusiasm. Linnaeus delineated the end of an era; he also heralded the beginning of modern botany.

If the earlier epochs were marked by great men, it is fair to say that after 1860, botany was marked by great ideas. Charles Darwin's theory of evolution by natural selection and Gregor Mendel's laws of heredity both had their impact on botany; from that time on there was a gradual linking of the divisions of descriptive, morphological, and functional botany. The rise of plant cytology, ecology, genetics, biochemistry, and other interrelated and interdependent botanical sciences began in the twentieth century, nurtured by the revolutions in biological thinking and encouraged by the success of federal funding, team research, and a myriad of new biotechnological techniques.

Botanical literature mirrors the development of the literature of other biological disciplines, which, in turn, can be characterized by the growth of the literature of science as a whole. The growth of scientific literature shows a gradual progression from the time of the ancients, through the Dark Ages, to the invention of the printing press by Gutenberg in the fifteenth century. Until that time scientific literature, composed mostly of herbals and descriptive writing, was not widely available. After the invention of the printing press, the printed word spread rapidly, with the first journal serving scientists, *Philosophical Transactions of the Royal Society of London*, established in 1665. During the past three centuries, there has been an exponential growth of scientific literature (de Solla Price, 1963). By 1800 there were 100 scientific journals, a number that had grown to an incredible 10,000 by 1900 and included such important botanical abstracting and indexing tools as *Botanisches Zentralblatt, Just's botanischer Jahresbericht*, and *Zeitschrift für Pflanzenkrankheiten* that helped keep the botanists of the day up-to-date in their reading. For a very interesting and informative discussion on the historical background of scientific communication in the life sciences, see Kronick (1985). In his book, Kronick traces the development of scientific literature from its oral roots to transmission by writing, printing, scholarly letters between scientific peers, newsletter columns, scientific journals, and societies.

Since the early 1900s, science and the literature that must accompany it have grown at an astonishing rate. Federally funded research, large research laboratories, world-wide scientific research programs, and numerous specialist journals have all produced a doubling of the periodical literature every 10 to 15 years during the twentieth century. Since World War II, scientific literature has grown logarithmically, with technological advances in electronic communication and storage having significant consequences on the future of scientific communication.

Biological literature can be described in various ways. Kronick (1985) bases his discussion on the distinguishing parameters of size and growth; distribution by country, subject, and language; scatter and use; obsolescence and redundancy; and writing and publishing cycles. The matter is further complicated by the discrepancies that occur between various studies conducted over time, and of course, while the characteristics of the botanical literature do conform overall to the qualities of scientific literature in general, it may be more meaningful to describe the botanical literature in a more specific manner.

One way to discuss botanical literature is to divide it into two parts: the descriptive and the functional (Lawrence, 1970). Descriptive botanical literature is cumulative and is of importance whenever and wherever it occurs, from the fifteenth century incunabula to the primary journals of the eighteenth, nineteenth, and twentieth centuries. It is a vast literature, which is worldwide in scope, multilingual in nature, and chronologically inclusive. It is expensive in terms of acquisition, space requirements, preservation, and conservation. Functional botanical literature, on the other hand, depends almost entirely on current serial literature for the most recent information on original research. Older materials are not as important for access; therefore, the problems of storage are not as demanding as those of descriptive literature. Functional literature may be found in a wide range of journals and can be very expensive to obtain due to its wide disciplinary scope and prolific rate of growth.

Information scientist and publisher Eugene Garfield has written several essays on the distinctive qualities of botanical literature. He stated that it has an annual growth rate of 3 percent, a doubling time of 25.5 years, and relies more on the older literature than a faster-moving field like biochemistry (Garfield, 1979). In general, botanists cite life and physical scientists, but the reverse is not true, although many fundamental discoveries in basic biological science were made by botanists (Garfield, 1981). Nonjournal publications, such as circulars and monographs, often play an important part in communication between botanical researchers and as many "classic" papers are published in multidisciplinary journals as in botany journals. In 1980, as much as 75 percent of the botanical literature reported basic research in plant physiology, an area which is paramount in botany at the present time. Additional analysis demonstrated that highly cited articles from plant physiology journals such as *Plant Physiology*, *Annual Review of Plant Physiology*, *Planta*, *Physiologia Plantarum*, and *Zeitschrift für Pflanzenphysiologie* ranged from an article published in 1949 to one published in 1975, with the mode designated by articles from the 1960s.

Surprising as it may seem, botany as a discipline is not represented among science's most-cited primary authors for 1961-75 (Garfield, 1980). Among the reasons for this phenomenon are the following: (1) many leading botanists were forced by the lack of funding for botany to leave the field for molecular or cell biology, leaving very few to teach future botanists; (2) botany is highly diversified, and such very specific areas may not be citing each other; (3) the primary author bias may be significant in botany if senior botanists list students as first author of significant papers; and (4) botany is not a basic science, but is a specialized field that builds on the basic sciences of chemistry and physics.

Be that as it may, in terms of recent research importance, the plant sciences today are certainly in the forefront of scientific advances, with biotechnology serving as a good example of the applied cutting edge. Biotechnology that makes use of plants has been of importance since antiquity—the rationale for exploitation of plant resources is old, but the techniques of deliberate control and manipulation of genetic materials are new. Since ancient times, the preservation of beverages by fermentation and of food by spices have been cottage industries that have made use of plant resources to help control food supplies. During the nineteenth century Industrial Revolution, these industries were expanded to factory scale, although the processes themselves changed very little. Near the end of the twentieth century, the development of botanical extractive processes and other industrial techniques to obtain products useful for health care and nutritional purposes has proven to be of major international economic and political importance. New biotechnological procedures of manipulating recombinant DNA and monoclonal antibodies have immense potential for plant biotechnology, with obvious applications for pharmaceuticals, agriculture, speciality chemicals and food additives, environmental control, energy production, and bioelectronics.

References

Cronquist, Arthur. 1981. *Basic botany*. 2nd ed. New York: Harper & Row.

de Solla Price, Derek J. 1963. *Little science, big science*. New York: Columbia University Press.

Garfield, Eugene. 1979. Trends in biochemical literature. *Trends in Biochemical Sciences* 4 (December): 290-95.

Garfield, Eugene. 1980. *Essays of an information scientist*. Vol. 3, 1977-78. Philadelphia: Institute for Scientific Information.

Garfield, Eugene. 1981. *Essays of an information scientist*. Vol. 4, 1979-80. Philadelphia: Institute for Scientific Information.

Kronick, David A. 1985. *The literature of the life sciences: Reading, writing, research.* Philadelphia: Institute for Scientific Information.

Lawrence, George H. M. 1970. Botanical libraries and collections. In *Encyclopedia of library and information sciences* 3, 104-21. New York: Marcel Dekker.

Parker, Sybil P. 1982. *Synopsis and classification of living organisms*. New York: McGraw-Hill.

PART I
BIBLIOGRAPHIC
CONTROL

1 Bibliographic Tools

Beginning with guides to the literature, this chapter provides an overview, a lead for the novice or uninitiated, to the library tools and information resources serving the botanical literature. It discusses those bibliographic tools, including guides to the literature, bibliographies, and catalogs of important botanical collections, that help verify and identify botanical materials. Emphasis here is on those materials that cover a wide chronological period or have a broad multifaceted scope; in other words, lists and information that cannot be reproduced easily using computerized databases. For the most part commercial, exhibition, and strictly regional catalogs or bibliographies are not included. Several of the annotated sources do not include price or ISBN information because of their age and unavailability. Nevertheless, they have been included because of their importance; if a botanist or librarian has access to these older materials, they may well be the first tool of choice as an information source.

There is an excellent discussion on the literature of science and technology in the fifth edition of the *McGraw-Hill Encyclopedia of Science and Technology* (1982). In this essay, George Bonn, the author of the article, outlines and defines the forms and characteristics of primary and secondary scientific literature and makes comparisons and distinctions between their functions and uses by scientists. Bonn defines a bibliography as a list of references to primary sources, arranged in various formats, covering the literature according to the bibliography's established parameters, and published separately or as part of a serial or larger monograph. Occasionally, bibliographies lead to biographical details about botanical authors, collectors, and explorers that are difficult to locate in the more traditionally historical sources, and so in some cases, this chapter will have relevance to chapter 8 which covers retrospective biographical materials. This is not an isolated case. Following the same rationale, it may be that the *Index to American Botanical Literature*, for example, would fit as well in chapter 2, covering indexes, as in this chapter. There is a great deal of similarity between the structure and function of this index and the structure and function of selected bibliographies or catalogs. The important thing to remember on a difficult search is not to allow arbitrary placement decisions and labels to get in the way—consult whatever library tools are necessary to solve a sticky problem. If bibliographies and catalogs do not yield the desired information, consider indexes, abstracts, or biographies as the question seems to warrant.

Following is a group of useful guides specifically designed to aid students, teachers, researchers, and librarians in finding their way in the botanical literature; bibliographies and catalogs complete the chapter.

Guides to the Literature

1. **Ainsworth and Bisby's Dictionary of the Fungi**. 7th ed. By D. L. Hawksworth, B. C. Sutton, and G. C. Ainsworth. Kew, England: Commonwealth Mycological Institute; distr., Forestburgh, N.Y.: Lubrecht and Cramer, 1983. 445p. ill. $27.50. ISBN 0851985157.

This dictionary is much more than the traditional dictionary; among other services, it provides an excellent, scholarly introduction to the literature of the fungi. The arrangement is alphabetical and includes mycological terms, authors' names, taxonomic divisions, and other more general terms, such as "literature," "colour," "ecology," and "methods." Under each of these entries, there is a concise, authoritative discussion with appropriate references to the important literature of the topic. Several pages are devoted to the "literature" entry describing textbooks, general taxonomic works, bibliographies, catalogs, floras, periodicals, etc., of significance in mycological work. This is the best guide available to the mycological literature; it covers current and retrospective primary and secondary literature sources, as well as providing information on the history of mycology and biographical notes on some well-known mycologists and lichenologists.

2. **American Reference Books Annual**. Vol. 1- . Edited by Bohdan S. Wynar. Littleton, Colo.: Libraries Unlimited, 1970- . annual. $70.00/yr. ISSN 0065-9959.

This title is a comprehensive review source for reference books in all areas, including the botanical sciences. All such books published or distributed in the United States each year are cited in full bibliographic entries followed by detailed descriptive and evaluative reviews written by authorities in the field.

3. Blanchard, J. Richard, and Lois Farrell. **Guide to Sources for Agricultural and Biological Research**. Berkeley, Calif.: University of California Press, 1981. 672p. $48.50. ISBN 0520032268.

Although the primary emphasis is on research in agriculture and biology, many related subjects such as botany, ecology, forestry, and meteorology are covered in detail in this comprehensive source. It has been called a "landmark agricultural reference tool" and as such, has relevance to botanical information. [R: ALIN, Mar. 1982, pp. 1-7]

4. Coffey, Janice C. "Soviet Journals Important for Taxonomic Botany: A Translation of Zaikonnikowa's List, with Emendations." **Huntia** 5, no. 2 (1984): 85-106.

This article updates, supplements, and corrects information in *Botanico-Periodicum-Huntianum*, concerning abbreviations for Soviet journals important for taxonomic reference.

5.   Crafts-Lighty, Anita. **Information Sources in Biotechnology**, 2nd ed. New York: Stockton Press, 1986. 320p. $100.00. ISBN 0943818184.

All aspects and divisions of biotechnological issues and information sources are covered in this very useful publication. Coverage includes monographs, conferences, trade periodicals, research and review periodicals, abstracting and secondary sources, computer databases, patents, market surveys, directories, organizations, and information services.

6.   Davis, Elisabeth B. **Using the Biological Literature: A Practical Guide**. New York: Marcel Dekker, 1981. 272p. bibliog. index. $45.00. ISBN 0824772091. (Books in Library and Information Science, Vol. 35).

Covers the reference works and computerized databases relevant to the basic biological sciences.

7.   **Guide to Reference Books**. 9th ed. Compiled by Eugene P. Sheehy. Chicago: American Library Association, 1976. 1015p. $40.00. ISBN 0838902057. **First Supplement:** 1980. 316p. $15.00pa. ISBN 0838902944. **Second Supplement:** 1982. 252p. $15.00pa. ISBN 0838903614.

Basic reference books for all areas of research are included with critical annotations and complete bibliographic information. Useful as a selection tool and as a finding aid for appropriate research resources, this valuable guide is especially helpful for locating retrospective materials.

8.   **Guide to Reference Material**. Vol. 1, **Science & Technology**. 4th ed. Edited by A. J. Walford. London: Library Association; distr., Chicago: American Library Association, 1980. 697p. $74.50. ISBN 0853656118. [Remainder of 4th ed. in preparation.]

Volume 1 is relevant to the literature of botany and includes many multidisciplinary sources of interest to botanists.

9.   Herner, Saul, with Gene P. Allen and Nancy D. Wright. **A Brief Guide to Sources of Scientific and Technical Information**. 2nd ed. Arlington, Va.: Information Resources, 1980. 160p. ill. $15.00pa. ISBN 0878150315.

Chapter 1, "Information Directories and Source Guidance," is particularly useful for biologists looking for "descriptive compendia of primary and secondary sources of information." Other chapters include "Major American Libraries and Resource Collections" and "Organization of Personal Index Files."

10.   Kronick, David A., assisted by Wendell D. Winters. **The Literature of the Life Sciences: Reading, Writing, Research**. Philadelphia: Institute for Scientific Information, 1985. 255p. bibliog. index. $29.95. ISBN 0894950452. (The Library and Information Science Series).

The aim of this book is to provide a general background of observations and useful information for the practitioner, investigator, and student in the life sciences, so that they will be able to read, write, and research the literature in a more intelligent and efficient manner. Kronick discusses his perspectives on the life sciences, their historical background, varieties of information sources, and characteristics. He writes about indexing, searching, building personal files, and communicating with other scientists. This is a philosophical, thoughtful overview that is relevant to botany as a discipline within the life sciences; Kronick provides a framework through which to view botany in a broad context.

11.   Lawrence, George Hill Mathewson. **Taxonomy of Vascular Plants**. New York: Macmillan, 1951. 823p. ill. ISBN 002368190X.

Chapter 14, "Literature of Taxonomic Botany," is a superior reference source for understanding and retrieving the retrospective literature of systematic botany. Chapter sections include annotations for indexes to taxonomic works, regional and national floras, dictionaries, maps, and other, older reference materials of interest in botanists. Supplementary information appears in "Botanical Libraries and Collections" by Lawrence, in *Encyclopedia of Library and Information Science* 3 (1970): 104-21.

12.   **Library Research Guide to Biology: Illustrated Search Strategy and Sources**. Compiled by Thomas G. Kirk, Jr. Ann Arbor, Mich.: Pierian Press, 1978. 84p. index. $19.50; $12.50pa. ISBN 087650098X; 0876500998pa. (Library Research Guides Series, No. 2).

This guide to research presents useful strategies for planning and conducting library research on biological topics. It is written for the undergraduate, but would be helpful to the interested layperson in the public library as well.

13.   Malinowsky, H. Robert, and Jeanne M. Richardson. **Science and Engineering Literature: A Guide to Reference Sources**. 3rd ed. Littleton, Colo.: Libraries Unlimited, 1980. 342p. index. $33.00; $21.00pa. ISBN 0872872300; 0872872459pa. (Library Science Text Series).

This is an excellent guide to the literature of the sciences. Botanical abstracts, indexes, bibliographies, encyclopedias, dictionaries, handbooks, field guides, treatises, and directories comprise about 15 pages of the volume.

14.   **Scientific and Technical Information Sources**. Edited by Ching-Chih Chen. Cambridge, Mass.: MIT Press, 1977. 519p. $50.00. ISBN 0262030624.

Covers all biological and technical reference sources, nonprint materials, computerized databases, and supplemental reading lists. Arrangement is by form of material. It is not comprehensive for botany.

15.   Smith, Roger C., W. Malcolm Reid, and Arlene E. Luchsinger. **Smith's Guide to the Literature of the Life Sciences**. 9th ed. Minneapolis, Minn.: Burgess, 1980. 223p. $15.95pa. ISBN 0808735764.

This guide discusses the literature of biology in general and devotes chapters to taxonomy, thesis preparation, scientific writing, how to search the literature, libraries and book classification, as well as annotations for the primary and secondary life sciences literature. Botany is included as a discipline and can be retrieved under the general headings for form or type of literature.

16.   Stieber, Michael T., et al. "Guide to the Botanical Records and Papers in the Archives of the Hunt Institute." Part I. **Huntia** 4, no. 1 (1981): 5-89. Part II. **Huntia** 5, no. 3 (1984): 129-210.

This guide describes the botanical materials in the archives of the Hunt Institute. Included are a detailed inventory of papers, letters, the oral history collection, and a cumulated index of names.

17. Swift, Lloyd H. **Botanical Bibliographies: A Guide to Bibliographic Materials Applicable to Botany**. Minneapolis, Minn.: Burgess, 1970; repr., 1974. 800p. $77.00. ISBN 387429076X.
This is an excellent guide to the bibliographic literature of botany and allied areas.

18. **The United States Government Manual 1984/85**. Office of the Federal Register, National Archives and Records Service, General Services Administration. Washington, D.C.: Government Printing Office, 1984. 913p. $12.00. ISBN 0318175878. GS4.109: 984-85.
This is the official handbook of the federal government; it is also issued as a special section of the *Federal Register*. The *Manual* provides comprehensive information on agencies of the legislative, judicial, and executive branches and is especially useful for activities/programs/publications. Each entry includes a list of principal officials, a summary statement of each particular agency's purpose and role in government, a brief history, a description of programs and activities, and a sources of information section with details on consumer activities, contracts and grants, employment, publications, and other areas of citizen interest.

19. **The Use of Biological Literature**. 2nd ed. Edited by R. T. Bottle and H. V. Wyatt. Hamden, Conn.: Archon, 1972. 384p. ill. ISBN 0408384115.
Although this guide is outdated, it still provides an excellent, comprehensive introduction to the literature of biology. Two chapters are devoted to botany and its literature; in addition there are pertinent discussions on the characteristics of scientific literature and libraries. Bottle and Wyatt have supplied important information on the literature of botanical taxonomy as well as an entry to retrospective works on the voluminous descriptive botanical resources.

Bibliographies and Catalogs

20. Andrews, Theodora, with the assistance of William L. Corya and Donald A. Stickel, Jr. **A Bibliography on Herbs, Herbal Medicine, "Natural" Foods, and Unconventional Medical Treatment**. Littleton, Colo.: Libraries Unlimited, 1982. 344p. illus. index. $30.00. ISBN 0872872882.
The title describes quite well this valuable compilation of 749 annotated cookbooks, periodicals, reference works, field guides, and popular and scientific monographs. The reviewer called this a "unique compilation that should prove useful in library settings." [R: RQ, Fall 1982, pp. 90-91]

21. **Annual Bibliography of the History of Natural History**. Vol. 1- . London: British Museum (Natural History), 1985- . annual. ISSN 0268-9936. (Vol. 1: 68p. $10.00. ISBN 0565009737).
Volume 1 of this comprehensive bibliography of articles and monographs relating to the history of natural history covers the literature of 1982 for biology, botany, zoology, entomology, paleontology, mineralogy, forestry, agriculture, geography, and explorations. Eight thousand periodicals are scanned to produce an alphabetical list

arranged by author; there are indexes by subject, biography, and institution. Volumes covering 1983 and 1984 should be published shortly; volumes after that should appear about six to eight months after the end of the year to which they refer.

22. Arber, Agnes. **Herbals: Their Origin and Evolution: A Chapter in the History of Botany 1470-1670**. 2nd ed. Cambridge, England: Cambridge University Press, 1953. 326p. ill. index.

Appendix 1 of this important contribution provides a chronological list of the principal herbals and related botanical works published between 1470 and 1670. The list is not exhaustive, and includes mostly only first editions. Appendix 2 is an alphabetical list of the historical and critical works consulted during the preparation of the book, and as such provides another valuable bibliography of interest to botanists. For more information, see entry 602.

23. Bay, Jens Christian. **Bibliographies of Botany: A Contribution toward a Bibliotheca Bibliographia**. Jena, Germany: Fischer, 1909. (In Progressus Rei Botanicae 3, no. 2 [1909]: 331-56).

"A valuable, annotated historical bibliography of bibliographies including detailed records of periodicals; general, local, and subject bibliographies; library catalogs; auction and sales catalogs, etc." [R: Winchell, 1967, p. 547]

24. **Bibliographical Contributions from the Lloyd Library**. Cincinnati, Ohio: Lloyd Library, 1911-18. 3v.

Basically a catalog of the Lloyd Library, this set contains: volume 1, a bibliography of periodical literature held in the Library, and a bibliography relating to floras of Europe, Great Britain, North and South America, Asia, and Africa; volume 2, catalog of periodical literature, a catalog of books and pamphlets of authors A-M; volume 3, catalog of books of authors N-Z. An updated version of the catalog of periodical literature in the Lloyd Library written by W. H. Aiken and S. Waldboot may be found in the *Bulletin of the Lloyd Library and Museum of Botany, Pharmacology, and Materia Medica*, No. 34 (Lloyd Library, 1936, 103p.).

25. **Bibliography of Systematic Mycology**. See entry 97.

26. Boivin, Bernard. "A Basic Bibliography of Botanical Biography and a Proposal for a More Elaborate Bibliography." **Taxon** 26, no. 1 (1977): 75-105.

Although Boivin's emphasis is on biographical sources (see entry 610), the bibliography provided in part 2 of this article is very useful for verification of authors, titles, and floras of particular regions.

27. **B-P-H (Botanico-Periodicum-Huntianum)**. Edited by George H. M. Lawrence, et al. Pittsburgh, Pa.: Hunt Botanical Library, 1968. 1063p. $24.00.

This very useful compendium provides information on all periodical titles that regularly contain articles dealing with plant sciences and botanical literature. Information includes abbreviations, full title, name of city where first volume was published, the sequence of volume numbers, and location of entry in the *Union List of Serials* when applicable.

28. **Catalogue of Botanical Books in the Collection of Rachel McMasters Miller Hunt**. Pittsburgh, Pa.: Hunt Botanical Library, 1958-61. 2v.

An interesting addition to this catalog is a historical introduction to botany, medical aspects of early botanical books, the beginnings of modern husbandry, and illustrations of early botanical works, all written by authorities in the field. Valuable information on sources and locations includes complete and detailed bibliographic descriptions of botanical books.

29. **Catalogue of the Farlow Reference Library of Cryptogamic Botany**. Boston: G. K. Hall, 1979. 6v. $660.00/set. ISBN 0816102791.

Sixty thousand books, periodicals, and reprints, many of them not held elsewhere in the United States, are represented in this collection of materials pertaining to non-flowering plants.

30. **Catalogue of the Library of the Academy of Natural Sciences of Philadelphia**. Boston: G. K. Hall, 1972. 16v. $1,530.00/set. ISBN 081610946X.

The Library, founded in 1815, is noted for its finely illustrated monographs, complete sets of early journals, long runs of serial publications of scientific societies, and publications of foreign geological surveys. Of special relevance to botany is the Library's strength in systematics.

31. **Catalogue of the Mycological Library of Howard A. Kelly**. Compiled by Louis C. C. Krieger. Baltimore, Md.: Privately printed, 1924. 260p.

This is a useful finding aid for mycological verification covering over 7,000 papers and books of the larger fungi, lichens, myxomycetes, and selected algae. Kelly was a physician, author, and mycologist with more than amateur interest in the subject; this catalog is compiled by a well-respected mycologist. The bibliography is arranged by author with additional lists of periodicals, floras, miscellanea, and addenda completing the volume.

32. **Catalogue of the Royal Botanic Gardens, Kew, England**. Boston: G. K. Hall, 1973. 9v. $495.00/author section; $415.00/classified section. ISBN 0816113807.

The Library of the Royal Botanic Gardens is particularly rich in early botanical books. Areas covered include plant taxonomy and distribution, economic botany, botanical travel and exploration, plant cytology, physiology, and biochemistry.

33. Desmond, Ray. **Bibliography of British Gardens**. Winchester, England: St. Paul's Bibliographies; distr., Charlottesville, Va.: University Press of Virginia, 1985. 224p. ill. bibliog. index. $30.00. ISBN 090679515X.

This guide to the literature, dealing with the descriptions of 5,500 individual gardens in England, Scotland, Wales, and Ireland, provides a historical survey, a guide to illustrations and photos, and suggestions for additional reading. There are county indexes for all included countries, plus the Channel Islands and the Isle of Man.

34. **Excerpta Botanica, Sectio B**. See entry 80.

35. Hall, Elizabeth C. **Printed Books 1481-1900 in the Horticultural Society of New York**. New York: Horticultural Society of New York, 1970. 279p.

This catalog lists 3,000 rare botanical and horticultural works in this exceptional collection. It is a short title catalog arranged by author with place and date of publication; the entries are not annotated. The scope of the collection is wide, to include incunabula, herbals, plant exploration, regional floras, botanical periodicals, botanical monographs by distinguished American botanists, botanical illustration, and early agricultural and horticultural works. Over 20 pages of reference materials are listed, arranged by subject, and there is a brief history of the Horticultural Society of New York.

36.   Haller, Albrecht von. **Bibliotheca Botanica**. Bern and Basel, Switzerland: n.p., 1771-72; repr., New York: Johnson, 1967. 2v. $110.00/set. ISBN 0384210503.

This very valuable bibliography provides bibliographical and biographical information for 1,500 primary botanical authors and 3,500 less important writers, illustrators, explorers, and collectors.

37.   Hawksworth, David Leslie, and Mark R. D. Seaward. **Lichenology in the British Isles, 1568-1975: A Historical and Bibliographical Survey**. Richmond, England: Richmond Publishing; distr., Eureka, Calif.: Mad River Press, 1977. 240p. ill. index. $65.95. ISBN 0916422321.

This is a comprehensive survey of lichenology in the British Isles including 2,695 entries covering books, papers, theses, and exsiccatae. The first part of the book deals with historical aspects; the bibliography is arranged by author and has a geographical index. The remaining portions of the volume are devoted to herbaria, collectors, herbarium abbreviations, exsiccatae, sources, and an index to the historical introduction. This is a very useful, scholarly book written by experts.

38.   Henrey, Blanche. **British Botanical and Horticultural Literature before 1800: Comprising a History and Bibliography of Botanical and Horticultural Books Printed in England, Scotland, and Ireland from the Earliest Times until 1800**. London: Oxford University Press; distr., New York: State Mutual Book, 1975. 3v. ill. indexes. $195.00/ set. ISBN 0686787390.

This is a work of scholarship that should stand the test of time very well; it is appropriate for large public and research libraries. It is annotated in entry 644; see also entry 44.

39.   **Index to American Botanical Literature 1886-1966**. Compiled by the Torrey Botanical Club, New York. Boston: G. K. Hall, 1969. 4v. $405.00/set. ISBN 0816113450.

This valuable listing of American botanical literature is arranged by author to include botanical literature from the Western Hemisphere. Subjects included are taxonomy, phylogeny, fungi floristics, bryophytes, pteridophytes, spermatophytes, morphology, anatomy, cytology, genetics, physiology, pathology, plant ecology, general botany, biography, and bibliography. Literature dealing exclusively with bacteriology, laboratory methods, manufactured products, and applied botany is not included. The first supplement was published in 1977 by G. K. Hall and comprised one volume ($110.00).

40.   **International Bibliography of Vegetation Maps**. Edited by August Wilhelm Kuchler and Jack McCormick. Naarden, Netherlands: Anton W. Van Bekhoven, 1965; repr., Lawrence, Kans.: University of Kansas Libraries, 1971. 4v. (University of Kansas Publications Library Series, 21).

Only published vegetation maps are listed in this bibliography, and data include title of the map, date of preparation, color, scale, legend in the original language if western European, author of the map, and date and place of publication. Arrangement is geographical and contents are: volume 1, vegetation maps of North America; volume

2, vegetation maps of Europe; volume 3, U.S.S.R., Asia, and Australia; volume 4, African, South American, and world maps.

41.   **International Catalogue of Scientific Literature. M: Botany**. London: Published for the International Council by the Royal Society of London, 1902-19; repr., New York: Johnson, n.d. 3v. $250.00/set. ISBN 0685232808.

This annual publication serves as a continuation to the *Catalogue of Scientific Papers*, and includes original, scientific botanical literature for the period 1901-14. The *M* section, serving as a botanical subject index to the worldwide scientific literature, is arranged by classification of morphology, anatomy, physiology, pathology, evolution, taxonomy, geographic distribution, and author. Each annual volume includes a list of journals scanned with their abbreviations. This is a unique source for the period covered that must be included in the collection of any botanical research library.

42.   Jackson, Benjamin D. **Guide to the Literature of Botany: Being a Classified Selection of Botanical Works, Including Nearly 6,000 Titles Not Given in Pritzel's** *Thesaurus*. Monticello, N.J.: Lubrecht and Cramer, n.d. 626p. $33.95. ISBN 3874290697.

This companion to Pritzel, a reprint of the 1881 edition, can be used for verification and as a finding aid for the older botanical literature. Nine thousand entries are arranged by subject classification; there is an index. This is an authoritative bibliography by a botanist/scholar and a "must" for any botanical research collection.

43.   Junk, Wilhelm. **Bibliographia Botanica**. Berlin: Junk, 1909. 1v. **Supplementum**. Berlin: Junk, 1916. 1v.

This is another valuable source for verifying older botanical materials. The bibliography of 6,891 botanical papers and books is arranged by topic in each volume and then by author. There are no indexes. Subjects include an important list of older botanical periodicals with complete bibliographic information, history, anatomy and physiology, taxonomy, economic plants, illustrations, plants arranged by broad division, and floras. Some entries are annotated in German.

44.   Kent, Douglas H. **Index to Botanical Monographs: A Guide to Monographs and Taxonomic Papers Relating Phanerogams and Vascular Cryptogams Found Growing Wild in the British Isles**. Published for the Botanical Society of the British Isles. New York: Academic, 1967. 163p. index. ISBN 0124044506.

Almost 1,900 references are included in this guide to papers dealing with botany, as described in the subtitle, relevant to the British Isles. The bibliography has a systematic arrangement under author with family and genera index access. This bibliography is useful for updating Jackson's *Guide to the Literature of Botany* (see entry 42); see *Key Works to the Fauna and Flora of the British Isles and Northwestern Europe* (entry 45) for supplemental information.

45.   **Key Works to the Fauna and Flora of the British Isles and Northwestern Europe**. Edited by G. J. Kerrich, D. L. Hawksworth, and R. S. Sims. New York: Academic, 1978. 179p. $40.00. ISBN 0124055508. (Systematics Association Special Volume, No. 9).

The purpose of this bibliography is to provide access to scientific books and papers that can be used to identify living organisms in the British Isles, surrounding seas, and other parts of northwest Europe. The volume has a taxonomic arrangement; fungi, plantae, and prokaryotes are covered in the botany section, which comprises approximately one third of the bibliography. Although the emphasis here is different than that in Kent's *Index to Botanical Monographs*, the latter may be used for additional materials (see entry 44).

46. Lindau, Gustav, and Paul Sydow. **Thesaurus Literaturae Mycologicae et Lichenologicae.** Lipsiis: Fratres Borntraeger, 1908-18. 5v. **Supplementum.** By R. Ciferri. Cortina, Italy: Papia, 1957-60. 4v. $300.00/set; $275.00/set(pa.). ISBN 0384327060; 0384327079pa.

These volumes list mycological books and papers published up to 1930. The original volumes 1-3 are author lists including 40,000 titles by 12,000 authors. Volumes 4-5 provide an analytical index to the first three volumes plus approximately 250 author biographies. The supplements list 31,000 additional titles by 6,000 authors. This is obviously a valuable retrospective research resource; it has been reprinted by Johnson Reprint Corporation (New York, 1954, 5v., $300.00/set).

47. Linne, Carl von. **A Catalogue of the Works of Linnaeus, Issued in Commemoration of the 250th Anniversary of the Birthday of Carolus Linnaeus, 1707-1778.** Stockholm: Sandbergs Bokhandel, 1957. 239p. (Sandbergs Antikvariatsfortechning, No. 12).

47a. Linne, Carl von. **Catalogue of the Works of Linnaeus (and Publications More Immediately Relating Thereto) Preserved in the Libraries of the British Museum (Bloomsbury) and the British Museum (Natural History) (South Kensington).** 2nd ed. London: British Museum, 1933. 246p. index.

These two catalogues are good examples of Linnaeus's work.

48. **Lynn Index, A Bibliography of Phytochemistry.** Vols. 1-5. Organized and edited by John W. Schermerhorn and Maynard W. Quimby. Boston: Massachusetts College of Pharmacy, 1957-62. Vol. 6. By Norman R. Farnsworth. Pittsburgh, Pa.: University of Pittsburgh, 1969. Vols. 7-8. By Norman R. Farnsworth. Chicago: University of Illinois at the Medical Center, 1972-74.

This bibliography, which covers the literature of phytochemistry from 1560, is based on the citation collection of Dr. Eldin V. Lynn. The materials are arranged taxonomically; each section includes descriptions of the botany, constituents, and references for each order or family. Volume 8 provides a cumulative index for orders and families.

49. Meisel, Max. **A Bibliography of American Natural History; The Pioneer Century, 1769-1865; The Role Played by the Scientific Societies; Scientific Journals; Natural History Museums and Botanic Gardens; State Geological and Natural History Surveys; Federal Exploring Expeditions in the Rise and Progress of American Botany, Geology, Mineralogy, Paleontology and Zoology.** New York: Premier Publishing, 1924-29. 3v.

These volumes aim to trace, bibliographically, the rise and progress of natural history in the United States; this record obviously has great relevance to the study of botany. Volume 1 is an annotated bibliography of the publications relating to history, biography, and bibliography of American natural history and its institutions from Colonial times through the pioneer century, published up to 1924. There are subject and geographic indexes and a selected bibliography of the biographies and bibliographies of the principal American naturalists of the time. Volumes 2 and 3 deal with the institutions that have contributed to American natural history from 1767 to 1865. Volume 3 also contains a bibliography of books, chronological tables, and an index of authors and institutions. There is a wealth of information in these volumes making them a worthy addition to reference and historical collections.

50. Merrill, Elmer Drew. "A Botanical Bibliography of the Islands of the Pacific." **Contributions from the United States National Herbarium** 30, no. 1 (1947): 1-322.

Almost 4,000 titles are briefly annotated in this bibliography of the Pacific Islands, including lower cryptogams, ferns, and seed plants in their economic, taxonomic, ecological, and other botanical aspects. The arrangement is alphabetical by author. This very useful bibliography by a giant in the field has subject and geographic indexes compiled by Egbert H. Walker, included on pages 323-404 in the same issue. To expand or update this bibliography, see Sachet and Fosberg's *Island Bibliographies* (entry 62).

51. Merrill, Elmer Drew, and Egbert H. Walker. **A Bibliography of Eastern Asiatic Botany**. Sponsored by the Smithsonian Institution, Arnold Arboretum, New York Botanical Garden, and Harvard-Yenching Institute. Jamaica Plain, Mass.: Arnold Arboretum of Harvard University, 1938. 719p. **Supplement I**. By Egbert H. Walker. Forestburgh, N.Y.: Lubrecht and Cramer, 1960. 552p. $18.50. ISBN 0934454116.

This is the most important work for this geographic area; it briefly annotates 21,000 papers and independently published materials. There is an excellent list of serial abbreviations, appendixes for older oriental works, and lists of oriental serials and authors. The subject index is geographical and systematic; there are also indexes for generic names and geographical names. The supplement brings the coverage up to 1958. These major bibliographies are appropriate for research libraries.

52. Miasek, Meryl A., and Charles R. Long. **Endangered Plant Species of the World and Their Endangered Habitats: A Compilation of the Literature**. rev. and enl. ed. New York: Council on Botanical and Horticultural Libraries, Library of the New York Botanical Garden, 1985. 153p. index. $2.50pa. (Plant Bibliography Series, No. 6).

Arranged by author within broad categories of general works, geographic districts, conservation and preservation, this valuable bibliography documents worldwide efforts to list endangered plant species and their special habitats. The bibliography attempts to be comprehensive to its publication date; it includes published and unpublished, national and international materials. All publications are available at the New York Botanical Garden. There is an addenda for very recent materials and an author index. This enlarged edition is much easier to use than the earlier one because of its arrangement allowing subject access.

53. Nissen, Claus. **Die botanische Buchillustration, ihre Geschichte und Bibliographie**. Stuttgart, Germany: Hiersemann, 1951-52. 2v.

Volume 1 is a history and volume 2 is a bibliography, by author, of 2,400 botanical books containing illustrations. There are indexes by artist, plant, country, and author.

54. Osler, William. **Bibliotheca Osleriana: A Catalogue of Books Illustrating the History of Medicine and Science, Collected, Arranged and Annotated by Sir William Osler**. Oxford, England: Clarendon Press, 1929. 875p.

Almost 8,000 annotated entries are arranged by topic with author, subject, and short title indexes. There are entries for herbals, pre-Linnean botany, medical botany, and materia medica. "Particularly valuable for its annotations." [R: Winchell, 1967, p. 601]

55. **Photosynthesis Bibliography**. Vol. 1- . Edited by Z. Sestak and J. Catsky. Dordrecht, Netherlands: Junk; distr., Hingham, Mass.: Kluwer/Boston, 1974- . annual. (Vol. 13: 1985. 412p. $65.00. ISBN 9061935334).

This bibliography includes papers in all fields of photosynthesis research. Arrangement is alphabetical by author; there are color-coded author, subject, and plant indexes in each volume. Coverage, beginning with volume 1, is from 1966 to date. This outstanding, comprehensive bibliography, included here because of its importance, is indispensable for researchers in photosynthesis and related work. [R: ARBA, 1981, entry 1417]

56. **Plant Science Catalog: Botany Subject Index of the U.S. National Agricultural Library**. Boston: G. K. Hall, 1958. 15v. $1,485.00/set. ISBN 0816105065.

This photographic reproduction of subject cards from the collection of the National Agricultural Library contains citations to world botanical literature from antiquity to 1952, including books, serials, proceedings, bulletins, textbooks, voyages, and biographies.

57. Pritzel, Georg August. **Thesaurus Literatureae Botanicae.** 2nd ed. Lipsiae: F. A. Brockhaus, 1872; repr., Monticello, N.Y.: Lubrecht and Cramer, 1972. 576p. indexes. $63.00. ISBN 3874290352.

This bibliography includes 11,000 titles by 3,000 authors and covers all botanical fields up to 1870. Pritzel and Jackson are two of the most important sources for information on separately published early botanical literature. Pritzel is supplemented by the Royal Society *Catalogue of Scientific Papers* (see entry 61).

58. **Recent Publications in Natural History.** Vol. 1- . New York: Department of Library Services, American Museum of Natural History, 1983- . quarterly. $10.00/yr. ISSN 0738-0925.

This bibliography of natural history is divided into 25 subject classifications to provide complete bibliographic information for recent publications. Many entries are annotated and a selected few receive full-length reviews. Citations for botanical materials are supplied by the staff of the Academy of Natural Sciences of Philadelphia.

59. Reed, Clyde F. **Bibliography to Floras of Southeast Asia: Burma, Laos, Thailand (Siam) Cambodia, Viet Nam (Tonkin, Annam, Cochinchina), Malay Peninsula, and Singapore**. Baltimore, Md.: Paul M. Harrod Co., 1969. 191p.

Although this bibliography is regional, it covers nineteenth- and twentieth-century publications that are often difficult to locate. It includes floristic works, taxonomic and monographic treatments of both vascular and nonvascular plants; cultivated, agricultural or medicinal plants; forestry, climatic phenomena relating to flora, and fossil flora. The arrangement is strictly alphabetical by author with no subject or geographical access by country, making it difficult to use if the preferred focus is narrower than Southeast Asia. For additional information, see Merrill and Walker's *A Bibliography of Eastern Asiatic Botany* (entry 51).

60.   Rehder, Alfred. **The Bradley Bibliography: A Guide to the Literature of the Woody Plants of the World, Published before the Beginning of the Twentieth Century**. Cambridge, Mass.: Riverside Press, 1911; repr., Monticello, N.Y.: Lubrecht and Cramer, 1976. 5v. $483.00/set. Vols. 1 and 2: ISBN 3874291073. Vol. 3: ISBN 3874291081. Vol. 4: ISBN 387429109X. Vol. 5: ISBN 3874291103.

The aim of this set is to include titles of all publications relating wholly or in part to woody plants. Books, pamphlets, and articles in serials in all languages are covered up to 1900. Volume 5 is an index to authors, subjects, and titles.

61.   Royal Society of London. **Catalogue of Scientific Papers. 1800-1900**. London: Clay, 1867-1902. 19v.

This is the most important listing of retrospective scientific papers for the period covered, the nineteenth century. The arrangement is by author, with reference to papers in 1,555 worldwide periodicals, European academies, and learned societies. It provides author's full name, periodical title and bibliographic citation for the article; abbreviations and list of journals scanned are included. Subject access is provided by a four-volume subject index (Cambridge University Press, 1908-14). For materials after 1900, see the *International Catalogue of Scientific Literature* (entry 41). The Royal Society *Catalogue* is a monumental piece of work that is an absolutely necessary part of the collection for any botanical research library.

62.   Sachet, Marie H., and F. R. Fosberg. **Island Bibliographies: Micronesian Botany; Land Environment and Ecology of Coral Atolls; Vegetation of Tropical Pacific Islands**. Compiled under the auspices of the Pacific Science Board. Washington, D.C.: National Academy of Sciences, National Research Council, 1955. 577p. (National Research Council Publication, 335).

Three separate annotated bibliographies, each with its own index, make up this volume. Merrill's great work, "A Botanical Bibliography of the Islands of the Pacific" (entry 50), forms the foundation for this subsequent publication on Pacific botany; a good research collection should contain both. A supplement, National Research Council Publication No. 1932, was published by the same authors in 1971 (Washington, D.C.: National Research Council, 427p.).

63.   Simon, James E., Alena F. Chadwick, and Lyle E. Craker. **Temperate Herbs: An Indexed Bibliography, 1971-1980: The Scientific Literature on Selected Herbs, and Aromatic and Medicinal Plants of the Temperate Zone**. Hamden, Conn.: Shoe String, 1984. 632p. index. $69.50. ISBN 0208019901.

This comprehensive bibliography on the major commercially significant herbs of the temperate zone was the 1985 winner of the Oberly Award for excellence in bibliographic literature from the Science and Technology Section of the Association of College and Research Libraries. References concern scientific details about the science of herbs. Part 1 is arranged by herb common name, part 2 presents literature arranged by subject, and part 3 lists references that include books, bibliographies, reports, conferences, and symposia.

64.   Swift, Lloyd H. **Botanical Bibliographies: A Guide to Bibliographic Materials Applicable to Botany**. Minneapolis, Minn.: Burgess, 1970; repr., 1974. 800p. $77.00. ISBN 387429076X.

This very useful guide is crammed with botanical bibliographies and should not be overlooked in the search for additional publications. It is also helpful for verification of titles and as a guide to the literature.

65.   **Water-in-Plants Bibliography**. Vol. 1- . The Hague: Junk, 1975- . annual. price varies. (Vol. 9: 1983. 187p. $46.50pa. ISBN 9061935202).

This bibliography is similar in arrangement to *Photosynthesis Bibliography* and contains papers in all fields of plant water relations research. It is international in scope and aims to publish citations to the more than 1,500 relevant papers dealing with plant water relations published each year. The bibliography is alphabetical by author with access through author, subject, and plant name indexes. Although it is specialized by topic, it is included here because of its value to botanists and to plant physiologists, in particular. [R: ARBA, 1984, entry 1303]

66.   Zanoni, Thomas, and Eileen Schofield. **Dyes from Plants: An Annotated List of References**. New York: Council on Botanical and Horticultural Libraries, Library of the New York Botanical Garden, 1983. $2.50pa. (Plant Bibliography Series, No. 5).

This useful listing of slightly over 60 titles includes books and articles with full bibliographic data and informative annotations discussing the use of plant substances in natural dyes.

2 Abstracts, Indexes, and Databases

Abstracting and indexing publications form an integral part of the secondary literature of botany by providing access to the primary literature of current and retrospective research journals. As discussed previously in the introduction, botanists were faced by the nineteenth century with the difficulty of keeping pace with an array of newly emerging journals. To alleviate this problem, abstracts and indexing tools were developed to manage the proliferating literature.

Because of the long history of the plant sciences and the structure of the botanical literature, this chapter will of necessity discuss many different publications covering the literature for access to articles appearing in scientific journals. Several of the tools annotated in the bibliography chapter, chapter 1, could as well be included here. For example, the chronological progression of the Royal Society *Catalogue*, the *International Catalogue of Scientific Literature*, and the various catalogs of famous botanical libraries do provide access to the scientific work reported before the modern era of computerized databases. Because botanists are faced with the task of going back into the literature over a period of several hundred years to answer questions of priority and taxonomy, the range of abstracting and indexing publications, along with their computerized counterparts, is wide and deep. As an aid to unraveling this complexity, this chapter discusses online access in combination with the particular printed tool that it complements, enhances, or replaces.

Abstracts and indexes are grouped together here because of their similar function of providing entry to primary literature; however, they are different in form and operation. Properly compiled, the index is probably the most useful of the two; convenient access to abstracts is dependent upon indexes for entry. Indexes can be produced in different arrangements to allow access by author, title, subject, geography, chronology, taxonomic category, or other categories. Both abstracts and indexes present bibliographic information for the research in question; however, abstracts include a summary of research not available in indexes. Because of this difference in information content, it is usually the case that indexes can be produced more rapidly than abstracts with a shorter lag time between journal publication and appearance in a secondary publication. As modern technology replaces older abstracting and indexing methods and production, this difference will no longer be valid.

Searching strategies for computerized databases and their printed versions are not examined here because access, command language, data fields, etc., for each tool vary to such a great extent. Suffice it to say that the literature searcher should consider carefully the instructions and manuals provided by the publisher and/or vendor for literature retrieval from either the printed publication or the online database.

Following is a list of the most important retrospective and current awareness abstracting and indexing publications for the botanical sciences. Because the emphasis is on domestic and English-language publications, two of the world's great abstracting services in the biological sciences are not included: the French *Bulletin signaletique*, and the Russian *Referativnyi Zhurnal*. Access to the French and Russian languages botanical literature is provided through the third comprehensive international abstracting source, *Biological Abstracts*, which is in English (see entry 69). Indexes to strictly taxonomic work are grouped together under the heading "Taxonomic Indexes," following, and separated from, the more general indexing and abstracting tools.

General Botanical Abstracts and Indexes

67. **Abstracts of Mycology**. Vol. 1- . Philadelphia: BioSciences Information Service, 1967- . monthly. $245.00/yr. including annual cumulative indexes. ISSN 0001-3617.

This abstract journal contains complete bibliographic data, abstracts and content summaries, in English, of research studies on fungi, lichens, and fungicides. The material covered is compiled from *Biological Abstracts* (entry 69) and *Biological Abstracts/RRM* (entry 70) to scan more than 9,000 serials and other publications from more than 100 countries. Four indexes provide access to author names, broad taxonomic categories, organism names, and specific words. Although the material covered is the same, the advantage of this offshoot from its two parent publications is that it provides a single reference source for worldwide research in mycology and its related areas of study. *Abstracts of Mycology* is available as part of BIOSIS PREVIEWS, a computerized literature database covering the literature from 1970 to date.

68. **Bibliography of Agriculture**. Vol. 1- . Phoenix, Ariz.: Oryx Press, 1942- . monthly. $895.00/yr. including annual cumulative indexes. ISSN 0006-1530.

Begun by the U.S. Department of Agriculture in 1942 and continued by Oryx Press, *Bibliography of Agriculture* indexes journal articles, pamphlets, government documents, special reports, and proceedings of importance in agriculture and allied sciences from data supplied by the Department of Agriculture. A controlled indexing vocabulary is used for the subject index; there is an author index. Online access is provided to the computerized database containing these files through AGRICOLA tapes supplied to vendors by the National Agricultural Library. AGRICOLA covers the literature from 1970 to date. The emphasis and focus of this index is primarily agricultural; however, it is important to botanists working in the applied area who need to pick up international information from government reports, agricultural experiment stations, and the like.

69. **Biological Abstracts.** Vol. 1- . Philadelphia: BioSciences Information Service, 1927- . semimonthly. $3,810.00/yr. (libraries), including cumulative semiannual indexes. ISSN 0006-3169.

This is the most comprehensive biological abstracting publication in the English language. It covers world literature including over 9,000 serial and nonserial publications; emphasis is placed on basic research papers from primary biological and biomedical journals. Access to abstracts is provided by four indexes: author, broad taxonomic categories, organism names, and specific subject words. The botanical sciences are covered in detail from every facet: general and systematic, ecological, physiological, biochemical, biophysical, and methodological.

Biological Abstracts is available as a computerized database, BIOSIS PREVIEWS, covering the literature from 1970 to date. The *Biosis Search Guide* is a reference manual designed to support users of the BIOSIS PREVIEWS machine-readable database; it is essential for conducting computerized searches and can be used as an index to vocabulary, a code book, and a text for use in developing effective strategies for literature retrieval. Back files of *BA* are available in microform from 1960.

BA, together with its companion publication *Biological Abstracts/RRM*, are the instruments of choice for comprehensive access to the literature; their scope and coverage provide the best single source available for current botanical research. The *Biosis List of Serials* gives an alphabetical list of the serials scanned for inclusion with their abbreviations and publisher's addresses. Beginning in 1985, documents reported in *BA* and *BA/RRM* are available from the University Microfilms International Article Clearinghouse in Ann Arbor, Michigan.

70. **Biological Abstracts/RRM (Reports, Reviews, Meetings).** Vol. 18- . Philadelphia: BioSciences Information Service, 1980- . monthly. $1,895.00/yr. (libraries), including semiannual cumulative indexes. ISSN 0192-6985.

This publication continues *Bioresearch Index* by the same publisher and may be considered a sister publication of *Biological Abstracts*. *BA/RRM* reports on notes, symposia papers, meeting abstracts, trade journal items, translated journals, review publications, bibliographies, technical data reports, research communications, books, book chapters, and a variety of special taxonomic publications. The literature monitored is the same as *Biological Abstracts*, and taken together, the two publications cover all aspects of worldwide botanical literature. *BA/RRM* is divided into three sections: (1) content summaries of reviews, reports and meetings; (2) synopses of books; and (3) meetings section of symposia and meeting papers. Access is provided by four indexes to author, taxonomic categories, organism names, and subject words. There are no abstracts. *BA/RRM* is available online as part of BIOSIS PREVIEWS covering the literature from 1970 to date. This computerized database allows instant combination of information from *BA* and *BA/RRM*, eliminating the necessity for scanning two publications in their printed versions. Online searches of BIOSIS PREVIEWS can be conducted most effectively with the *Biosis Search Guide*, mentioned with *Biological Abstracts* (entry 69).

71. **Biological and Agricultural Index**. Vol. 1- . New York: H. W. Wilson, 1916- . monthly, with annual cumulations. service basis. Back volumes, $100.00/yr. ISSN 0006-3177.

Scanning 204 English language periodicals, this index covers all aspects of biology and agriculture, including botany. There is a subject index with extensive cross-references; complete bibliographic information is provided, with an author index. The level of this index is appropriate for the undergraduate student or for the general public. Online access to *Biological and Agricultural Index* is available through WILSONLINE, the Wilson Company's online retrieval service.

72. **Biology Digest**. Vol. 1- . Medford, N.J.: Plexus, 1974- . monthly, September through May. $117.50/yr. including cumulative index. ISSN 0095-2958.

This abstracting publication covers 200 periodicals in biology, including botany. It was created to meet the needs of high school and undergraduate students, but it is also useful for amateur and professional botanists because it covers some of the more popular literature not included in *Biological Abstracts*. It is a competitor to *Biological and Agricultural Index* (entry 71), but unlike the *Index*, it provides comprehensible, lengthy abstracts. Each issue contains a subject and author index, plus a feature article on some aspect of the biological sciences.

73. **Biotechnology Research Abstracts**. Vol. 1- . Bethesda, Md.: Cambridge Scientific Abstracts, 1984- . monthly. $414.00/yr. ISSN 0733-5709.

This speciality abstracting tool provides abstracts for key articles from over 5,000 scientific journals covering genetic engineering, immobilization, cell culture, products of biotechnology, applications, fermentation, process engineering, patents, etc. Complete bibliographic information is provided for each abstract; subject and author indexes are provided for each issue with annual cumulations. This database is available online as part of the LIFE SCIENCES COLLECTION. For more information on abstracting and indexing services specifically geared to biotechnology, see *Information Sources in Biotechnology* (entry 5).

74. **Botanical Abstracts**. Vols. 1-15. Baltimore, Md.: Williams and Wilkins, 1918-26.

This was an abstracting serial with an international scope. It was continued and expanded to all of the biological sciences by *Biological Abstracts*. These abstracts are useful retrospectively.

75. **Botanisches Zentralblatt: Referiendes Organ für das Gesamtgebiet der Botanik**. Vols. 1-179. Im Auftrage der Deutschen Botanischen Gesellschaft. Jena, Germany: Fischer, 1880-1945.

This defunct German abstracting tool for botany is included here because of its coverage for retrospective materials. It picks up from the Royal Society *Catalogue* (entry 61) and continues through the period of the *International Catalogue of Scientific Literature* (entry 41) through the less than completely scanned literature covered by *Botanical Abstracts* (entry 74) and the fledgling *Biological Abstracts* (entry 69).

76. **Chemical Abstracts**. Vol. 1- . Columbus, Ohio: Chemical Abstracts Service, 1907- . weekly. $7,800.00/yr. (libraries), including indexes. ISSN 0009-2258.

This is the most comprehensive abstracting service for chemistry in English. It is available separately by section and online as CA SEARCH from 1970 to date. Although it does not cover botany specifically, it does cover botanical subjects from a chemical viewpoint and is absolutely essential for information on chemical formulas, molecular weights, chemical registry numbers, structural formulas, patents, and other biochemical information. *CA* is international in scope and provides a list of serials scanned, *Chemical Abstracts Service Source Index (CASSI)*, that is a gold mine of information on serials and continuing congresses pertinent to the chemical sciences. Chemical Abstracts Service operates a document delivery service for all materials abstracted in *CA*.

77. **Current Advances in Plant Science (CAPS)**. Vol. 1- . Elmsford, N.Y.: Pergamon Press, 1972- . monthly. $350.00/yr. ISSN 0306-4484.

This current awareness service for over 3,000 biological journals covers all aspects of the plant sciences. Arranged by subject, complete bibliographic information is available for articles scanned; there are no abstracts. This competitor for *Biological Abstracts* is not as comprehensive as *BA*; it does not provide as much information; and it is less expensive. The technical level of the two publications is equivalent and is aimed at the professional botanist, researcher, and upper-level college student. *CAPS* is part of the CURRENT AWARENESS IN BIOLOGICAL SCIENCES (CABS) database produced by Pergamon and is available online from that publisher. There is a document delivery service for CABS.

78. **Current Contents/Agricultural, Biology, and Environmental Sciences**. Vol. 1- . Philadelphia: Institute for Scientific Information, 1970- . weekly. $272.00/yr. ISSN 0011-3379.

79. **Current Contents/Life Sciences**. Vol. 1- . Philadelphia: Institute for Scientific Information, 1958- . weekly. $272.00/yr. ISSN 0011-3409.

These two sections of the *Current Contents* publications provide a unique service that alerts professional botanists to journal articles and books at time of publication. Each of these serials consists of the contents pages from over 1,000 selected, important periodicals relevant to their subject coverage as delineated by their title. This alerting service is useful in filling the two- to nine-month gap in coverage left between the citation of research articles in abstracting and indexing publications and the appearance of journal articles just off the press. Each issue of *Current Contents* contains a subject and author index, with current addresses of the author and a publisher's directory. There is a document delivery service available for articles listed in *CC*. These publications are appropriate for university and college collections. *Current Contents* is the current awareness tool of choice for keeping up with the literature because of its currency, frequency, convenience, and efficiency of arrangement.

80. **Excerpta Botanica**. Vol. 1- . New York: Fischer, 1959- . Edited in connection with the International Association for Plant Taxonomy, Utrecht, Netherlands. **Sectio A: Taxonomica et Chlorologica**. irregular. $260.00/yr. ISSN 0014-4037. **Sectio B: Sociologica**. quarterly. $89.00/yr. ISSN 0014-4045.

Section A provides abstracts in English, French, German, and Spanish for international coverage of articles on systematic botany, herbaria, gardens, etc. Section B presents a world bibliography of books and articles dealing with plant geography and ecology. These are important publications because of the auspices under which they are produced. A comprehensive search of the botanical literature should include them.

81. **Government Reports Announcements and Index**. Vol. 75- . Washington, D.C.: National Technical Information Services, U.S. Department of Commerce, 1975- . biweekly. $379.00/yr. ISSN 0097-9007.

Formerly *U.S. Government Research and Development Reports* and *Government Reports Announcement* (1964-74), this publication is a listing of technical reports from NTIS. It is arranged by subject, covering technology, engineering, and the physical, life, and social sciences. *GPAI* cites unclassified government-sponsored research, development, and engineering reports and analyses by government agencies, contractors, or grantees. Unique to other services listed here, *GPAI* reports unpublished material originating outside the United States. Information includes full bibliographic details and abstracts of work accomplished. Indexes include corporate author, subject, personal author, contract number, and accession/report number. The database is available online as NTIS, covering 1964 to date.

82. **Index Holmensis**. Edited by Hans Tralau. Vol. 1- . Zurich, Switzerland: Scientific Publishers, 1969- . irregular. price unavailable.

This world index of plant distribution maps provides complete bibliographic data and information on area covered. The volumes are arranged alphabetically by genus, species, and plant name. Volume contents are: volume 1: Equisetales, Isoetales, Lycopodiales, Psilotales, Filicales, Gymnospermae; volume 2: Monocotyledoneae, A-I; volume 3: Monocotyledoneae, J-Z; volume 4: Dicotyledoneae, A-B; and volume 5: Dicotyledoneae, C.

83. **Index to American Botanical Literature, 1886-1966**. Compiled by the Torrey Botanical Club, New York. Boston: G. K. Hall, 1969. 4v. $405.00/set. This invaluable bibliographic index is annotated in entry 39.

84. **Index to Illustrations of Living Things Outside of North America: Where to Find Pictures of Flora and Fauna**. Compiled by Lucile Thompson Munz and Nedra G. Slauson. Hamden, Conn.: Shoe String, 1981. 441p. ill. bibliog. index. $49.50. ISBN 0208018573.

This and entry 85 are companion volumes and indexes to illustrations of plants and animals throughout the world. Their formats are identical. This volume includes sources found in guides, handbooks, and encyclopedias dating mainly from 1963 through the 1970s. Both of these indexes are recommended for libraries of all sizes; research libraries will need to augment this information with more comprehensive sources (see entries, 99, 103, and 104). [R: RQ, Fall 1982, pp. 93-94]

85. **Index to Illustrations of the Natural World: Where to Find Pictures of the Living Things of North America**. Compiled by John W. Thompson. Edited by Nedra G. Slauson. Syracuse, N.Y.: Gaylord, 1977; repr., Hamden, Conn.: Shoe String, 1983. 256p. ill. bibliog. index. $42.50. ISBN 0208020381.

This index provides an alphabetical listing of 6,200 North American plants, birds, and animals with citations to where illustrations may be found. There is a scientific name index and a bibliography of sources. This very useful compilation is recommended for all libraries, particularly small public ones; the sources were chosen to be widely available with illustrations that clearly identify. Useful for the more popular plants, this index cannot compete, of course, with the more comprehensive and retrospective sources *Index Kewensis* (entry 103), Nissen's *Die botanische Buchillustration* (entry 53), *Index Londinensis* (entry 104), and the *Flowering Plant Index of Illustration and Information* (entry 99). For information concerning more worldwide taxonomic indexes for plant illustrations, see the section, "Taxonomic Indexes," in this chapter. [R: ARBA, 1978, entry 1277]

86. **Index to Plant Chromosome Numbers, 1979-1981**. Edited by Peter Goldblatt. St. Louis, Mo.: Missouri Botanical Garden, 1984. 427p. $15.00. ISSN 06161-1542. (Monographs in Systematic Botany, Vol. 8).

This is a continuation of the previous *Index to Plant Chromosome Numbers*, which covered 1975-78, and of earlier volumes in the Regnum Vegetabile series published by Bohn, Scheltema and Holkema, volumes 50, 55, 59, 68, 77, 84, 90, 91, 96. To provide references to the literature of plant chromosome number information for the period covered, entries are arranged alphabetically by family within these groupings: algae, fungi, bryophytes, pteridophytes, and spermatophytes. These compilations provide a useful service for botanists and should be included in any botanical library; between compilations, chromosome number information is updated in the periodical, *Taxon*.

87. **Index to Scientific and Technical Proceedings**. Vol. 1- . Philadelphia: Institute for Scientific Information, 1978- . monthly, with semiannual cumulations. $775.00/yr. ISSN 0149-8088.

This multidisciplinary series is designed to verify publication of scientific conference proceedings. It is appropriate for retrospective and current searches, providing complete bibliographic information. Both conferences and individual papers are described with indexes for category, subject, author/editor, sponsor, corporate entity, and meeting location. Over 3,800 journals are monitored, with coverage approximately 35 percent for life sciences, 10 percent for agricultural and biological sciences, 35 percent for engineering, and 20 percent for physical and chemical sciences. As of 1985, *Biological Abstracts/RRM* (entry 70) has greatly increased its coverage of symposia, conferences, and meetings. The advantages of *BA/RRM* besides its comprehensiveness are the extensive indexing and content summaries with taxonomic information that are of particular concern to the botanical sciences.

88. **Index to Scientific Reviews**. Vol. 1- . Philadelphia: Institute for Scientific Reviews, 1974- . semiannual, with annual cumulations. $550.00/yr. ISSN 0360-0661.

This publication assists in bibliographic control of the review literature. It is an international interdisciplinary index for science, medicine, agriculture, technology, and the behavioral sciences and includes all articles from the *Science Citation Index* (entry 95) that have 40 or more references, that are coded as review articles, and that include "Advances," "Progress," "Review," etc., in their titles. Citation, subject, author, and sponsoring organization indexes are provided for access. While this is a very useful index, it is not as comprehensive for review articles as *Biological Abstracts* (entry 69), for example, and serves botany as only a portion of its total coverage.

89. **Just's botanischer Jahresbericht: Systematisch geordnetes Repertorium der botanischen Literatur aller Länder**. Berlin-Zehlendorf: Borntraeger, 1873-1944. 63v.

This annual bibliography of world botanical literature was one of the first library tools developed to aid in keeping up with the swift growth of primary journals during the 1800s. Some abstracts are included, with annual indexes for authors and taxa. *Just's* is useful for retrospective searches and to fill the gap between the *International Catalogue of Scientific Literature* (entry 41) and *Botanical Abstracts* (entry 74).

90. Kent, Douglas H. **Index to Botanical Monographs: A Guide to Monographs and Taxonomic Papers Relating to Phanerogams and Vascular Cryptogams Found Growing Wild in the British Isles.** New York: Published for the Botanical Society of the British Isles by Academic Press, 1967. 163p. ISBN 0124044506.

This index is useful for retrospective work relevant to flora of the British Isles published since 1800. It includes monographs and taxonomic papers in a systematic arrangement and a list of abbreviations of the titles of periodicals of the same period. It complements and updates Henrey (entry 38), who covered the monographic botanical literature of Great Britain appearing before 1800.

91. **Microbiology Abstracts, Section C: Algology, Mycology and Protozoology.** Vol. 1- . London: Information Retrieval Ltd., 1972- . monthly. $465.00/yr. ISSN 0301-2328.

Continued in Bethesda, Maryland by Cambridge Scientific Abstracts, this abstracting publication scans over 5,000 scientific journals and other sources to include information on algae, fungi, lichens, and protozoa. It is arranged by subject and also includes book notices and announcements of proceedings, with author and subject indexes. *Microbiology Abstracts* is available online as part of the LIFE SCIENCES COLLECTION providing access to the literature from 1978 to date. This publication is not as comprehensive as *Biological Abstracts* (entry 69), but it does include unique references not found in *BA* and *BA/RRM*, so a comprehensive search of the literature for algae, fungi, and lichens should include it.

92. **Monthly Catalog of United States Government Publications.** Vol. 1- . Washington, D.C.: Government Printing Office, 1895- . monthly. $217.00/yr. ISSN 0362-6830.

This well-known publication is essential for access to government publications from the popular to the more technical, making it an indexing periodical appropriate for public, academic, and research libraries. For example, scientific reports issued by the U.S. Department of Agriculture are listed in this multidisciplinary tool, which covers journal articles, books, bibliographies, indexes, maps, government hearings, laws, executive orders, proclamations, and treatises in agriculture, business, law, medicine, engineering, sciences, social sciences, and humanities. Arrangement is by government department with author, title, subject, and series/report indexes in each issue; there are semiannual indexes and an annual cumulation. Complete bibliographic information is supplied for each entry including price and ordering information. GPO MONTHLY CATALOG is available as a computerized database reporting the government literature from 1976 to date.

93. Reuss, Jeremias David. **Repertorium Commentationum a Societatibus Litterariis Editarum. Secundum Disciplinarum Ordinem Digessit I. D. Reuss.** . . . . Gottingae: Dieterich, 1801-21; repr., New York: Burt Franklin; distr., New York: Lenox Hill, 1961. 16v. $550.00. ISBN 0833729667.

Useful for retrospective work, this index covers publications of learned societies to 1800, making it a precursor to the Royal Society *Catalogue* (entry 61). Volume 2 covers botany and mineralogy.

94. **Review of Plant Pathology**. Vol. 1- . Slough, England: Commonwealth Agricultural Bureaux, 1922- . monthly. $174.76/yr. ISSN 0034-6438.

Although this publication is somewhat out-of-scope, it is included because of its importance and relevance for mycologists and botanical microbiologists. Abstracts and review articles of importance to plant diseases, decay, fungicides, general and systematic mycology are included in the *Review*. Coverage is international with author and subject indexes in each issue, cumulated annually. Abstracts are arranged by subject with complete bibliographic information provided. Literature from developing countries is covered more completely here than in *Biological Abstracts* (entry 69). For a comprehensive search on issues dealing with plant pathology, this periodical should not be overlooked. It is available online as CAB (Commonwealth Agricultural Bureaux).

95. **Science Citation Index**. Vol. 1- . Philadelphia: Institute for Scientific Information, 1961- . bimonthly. price varies. ISSN 0036-827X.

This unique index provides information on citations to the scientific literature, who cites whom, and what is cited. It is multidisciplinary in nature and covers approximately 3,000 of the most influential international journals. Author (source), subject, corporate, and citation indexes are included in each issue, with annual and five-year cumulations available. Although *SCI* is expensive, it is one-of-a-kind and must be considered a most useful and original index. The online version, SCISEARCH, covers the literature from 1974 to date, scanning the *Current Contents* (entries 78 and 79) journals in addition to the journals in the printed version of *SCI*, for a total of over 4,000 journals. This larger coverage gives the online database a handy advantage over its printed counterpart.

96. **Virology Abstracts**. Vol. 1- . London: Information Retrieval Ltd., 1967- . monthly. $463.00/yr. ISSN 0042-6830.

This is continued in Bethesda, Maryland by Cambridge Scientific Abstracts. With relevance to plant virologists, this abstracting service is available online as the LIFE SCIENCES COLLECTION covering 1974 to date. Scope is international; abstracts retrieved from over 5,000 journals are arranged by subject, with author and subject indexes, and annual cumulations.

Taxonomic Indexes

Taxonomic indexes can be used as sources for the original publication of a plant name, for the classification (order, family, subfamily, or genus) to which a particular plant has been assigned, and for determining the correct name and spelling for the botanical and/or common name or synonym for a plant. To verify a plant name, it may be necessary to use several sources not only in this chapter, but also from other chapters. For example, entry 283 is an especially useful reference for identifying plant names and should be considered even though it is not included, perhaps somewhat arbitrarily, with the "Taxonomic Indexes." Other places to look for valid plant names are in the "Identification Sources" in chapter 7.

97. **Bibliography of Systematic Mycology**. Vol. 1- . Slough, England: Commonwealth Agricultural Bureaux. 1943- . semiannual. $25.00. ISSN 0006-1573.

This listing of world literature for systematic mycology is the only current aware-ness publication entirely concerned with the identification and taxonomy of fungi. Citations are arranged by systematic group and there is an author index in each issue; book reviews are also included. In 1984 the *Bibliography* was adjusted to bring the taxonomic information into compliance with *Ainsworth and Bisby's Dictionary* (see entry 1).

98. **Biosystematic Literature: Contributions to a Biosystematic Literature Index (1945-1964)**. Edited by O. T. and T. W. J. Gadella. Utrecht, Netherlands: International Bureau for Plant Taxonomy and Nomenclature, Tweede Transitorium, 1970. 566p. (Regnum Vegetabile, Vol. 69).

This is a source of references for use by people without the benefit of a large library at hand. It is not intended to be comprehensive, although it is international in scope for the years covered.

99. **Flowering Plant Index of Illustration and Information**. Compiled by R. T. Isaacson. Boston: G. K. Hall, 1979. 3v. $210.00/set. ISBN 0816103011. **First Supple-ment: 1979-81**. Boston: G. K. Hall, 1982. 2v. $235.00/set. ISBN 0816104034.

This set is a useful source for locating colored illustrations of flowering plants and can be used to update *Index Londinensis* (entry 104) and *Index Kewensis* (entry 103). It is appropriate for all botanical and large public libraries. There are cross-references for common and botanical names. See also entry 53 for a bibliography of books containing illustrative matter.

100. **Gray Herbarium Index: Reproduction of a Card Index, Issued 1894-1903**. Harvard University. Boston: G. K. Hall, 1968. 10v. **Supplement 1- . 1978- .**

This is an index to the names of flowering plants of the Western Hemisphere with citations to their authority and publication history. It duplicates *Index Kewensis* (entry 103) in part; it is especially useful for verifying the names of New World plants. Commencing with the equivalent of issues 304-306, the *Index* will be incorporated into a database and produced on microfiche.

101. **Index Filicum. Supplementum Quintum: pro annis: 1961-1975**. By F. M. Jarrett, et al. Oxford, England: Oxford University Press, 1985. 400p. $39.95. ISBN 0198545797.

This lists in alphabetical order the names of ferns and fern allies from family to species published from 1961-75. It is the latest volume in the original series, *Index Filicum*, that was begun in 1906 by C. F. Christensen. It is to ferns what *Index Kewensis* is to the taxonomy of flowering plants.

102. **Index Hepaticarum**. Vol. 1- . Vaduz, Liechtenstein: J. Cramer. 1962- . Vol. 10: 1985. 352p. $52.50. ISBN 3768211002.

This index presents nomenclatural, bibliographical, and systematic data for Hepaticae. It serves the same sort of function as does *Index Kewensis* (entry 103), *Index Filicum* (entry 101), and *Sylloge Fungorum* (entry 116), and is updated in *Taxon* (entry 193).

103. **Index Kewensis Plantarum Phanerogamarum Nomina et Synonyma Omnium Generum et Specierum a Linnaeo usque ad annum MDCCCLXXXV Complectens Nomine Recepto Auctore Patria Unicuique Plantae Subjectis.** Vols. 1-2. Sumptibus beati Caroli Roberti Darwin dectu et consilio Josephi D. Hooker confecit B. Dayton Jackson. Oxford, England: Clarendon Press, 1893-95; repr., Forestburgh, N.Y.: Lubrecht and Cramer, 1977. $465.00/set. ISBN 3874291170. **Supplementum**. 1886- . Vol. 1- .

Supplements are issued every five years. Supplement 16 (1981) covers 1971-76. This is an indispensable reference. Although the earlier volumes, particularly volumes 1 and 2 of the original set, contain many mistakes, the later supplements are reputed to be much more accurate. There is a cumulated microfiche index through supplement 16. Entries marked by an asterisk, beginning with supplement 10, indicate articles with illustrations.

104. **Index Londinensis to Illustrations of Flowering Plants, Ferns and Fern Allies: Being an Emended and Enlarged Edition Continued Up to the End of the Year 1920 of Pritzel's Alphabetical Register of Representations of Flowering Plants and Ferns Compiled from Botanical and Horticultural Publications of the Eighteenth and Nineteenth Centuries.** By O. Stapfanel and W. C. Wordsell. Kew, England: Royal Horticultural Society of London, 1929-31. 6v. **Supplement** (for the years 1921-35). Oxford, England: Clarendon Press, 1941; repr., Forestburgh, N.Y.: Lubrecht and Cramer, 1979. $1,008.00/set. ISBN 3839100000.

This index is continued by *Index Kewensis* (entry 103) supplements and can be updated by the *Flowering Plant Index* (entry 99). Also, see entry 53 for an index to plant illustrations.

105. **Index Muscorum**. R. Van Der Wijk, chief editor. Utrecht, Netherlands: International Bureau for Plant Taxonomy and Nomenclature, 1959-69. 5v. (Regnum Vegetabile, Vols. 17, 26, 33, 48, 65).

This is a compendium setting forth the nomenclatural status of the *Musci*, until 1969. There is information concerning published names, valid or not, substitute names, and new combinations of names for mosses. Updated in *The Bryologist* (entry 152).

106. **Index Nominum Genericorum (Plantarum)**. Edited by Ellen R. Farr, Jan H. Leussink, and Frans A. Stafleu. Utrecht, Netherlands: International Bureau for Plant Taxonomy and Nomenclature, 1979. 3v. $315.00/set. ISBN 9031303272. (Regnum Vegetabile, Vols. 100-103).

This is a list of validly published scientific plant names of all genera, recent and fossil, and includes citations to authors, references to place and time of publication, homonymy, indications of taxonomic placement, and additional information on names.

107. **Index of Fungi**. Vol. 1- . Kew, England: Commonwealth Mycological Institute, 1940- . semiannual. $40.00/yr. ISSN 0019-3895.

This index provides "a list of names of new genera, species and varieties of fungi, new combinations and new names, compiled from world literature." This standard reference work contains 10-year cumulative indexes and some supplements. It supersedes Petrak's lists (entry 114) to provide full bibliographic citations for fungi and lichens since 1971.

108. **Index to Grass Species**. Smithsonian Institution, compiled by Agnes Chase and C. D. Niles. Boston: G. K. Hall, 1963. 3v. $300.00/set. ISBN 081610445X.

This index is a reproduction of a card index kept by the U.S. Department of Agriculture for many years. It is international in scope and lists names of the species of grasses described from 1763-1962. Information provided includes scientific name, authority with bibliographic citation to the publication in which it appeared, type locality, and country. This is useful in the same way as *Index Kewensis* (entry 103), although it covers a different part of the plant kingdom.

109. **Index to Plant Distribution Maps in North American Periodicals through 1972**. Compiled by W. Louis Phillips and Ronald L. Stuckey. Boston: G. K. Hall, 1976. 686p. $115.00. ISBN 0816100098.

There are 28,500 entries arranged alphabetically by taxa to represent 268 periodicals. This index lists mapped taxa to provide citations to periodicals giving information on geographical distribution, type of map, and author of the article. This is a useful source for plant taxonomy, ecology, and biogeography.

110. Kent, Douglas H. See entry 44 for another useful taxonomic index.

111. **Kew Record of Taxonomic Literature**. London: Her Majesty's Stationery Office, 1971- . annual. (1985. 526p. $150.00. ISBN 0112411703).

This 1985 publication of 1980 literature reports worldwide taxonomic literature of the flowering plants, gymnosperms, and ferns in a systematic arrangement. It is comprehensive in scope to include all articles, books, papers, and all new names with the exception of cultivars. Entries are arranged systematically whenever possible with more general papers listed in a subject arrangement to include bibliography, botanical institutions, floristics, etc. An author index, geographic information, and abbreviations for periodical titles are included. Lag time is approximately four years for this publication.

112. Love, Askell, and Doris Love. **Cytotaxonomical Atlas of Arctic Flora**. Vaduz, Liechtenstein: J. Cramer, 1975. ill. bibliog. index. 598p. $80.00. ISBN 3768209768.

The purpose of this atlas is to serve as a checklist of the families, genera, species, and subspecies of vascular plants that occur naturally in the northlands. It is also a critical review of the chromosome numbers of the taxa included. After a brief introduction, the atlas is arranged taxonomically providing chromosome information with citations to the literature that contained the original information, synonyms, and longitudinal and latitudinal distribution for the species. Love and Love are respected for their work with chromosome numbers and for their synthesis of difficult-to-find information. Three other books by the same authors are relevant to this topic: *Cytotaxonomical Atlas of the Pteridophyta* (J. Cramer, 1977); *Cytotaxonomical Atlas of the Slovenian Flora* (Lubrecht and Cramer, 1974); and a text, *Plant Chromosomes* (Lubrecht and Cramer, 1975). This information is updated in *Taxon*.

113. **Mycotaxon**. See entry 169.

114. Petrak, Franz. **List of New Species and Varieties of Fungi, New Combinations and New Names Published, 1920-1939**. Kew, England: Commonwealth Mycological Institute, 1950-57. $115.75/set.

This work covers the mycological literature from 1922 through 1935. It is superseded by *Index of Fungi* (entry 107). *An Annotated Index to the Mycological Writings of Franz Petrak*, compiled by Gary J. Samuels, has been published since 1981 by the New Zealand Department of Scientific and Industrial Research, Wellington (Vol. 4: 1985, $12.50, ISBN 0477067611, *DSIR Bulletin*, 230).

115. Rehder. Alfred. **Bibliography of Cultivated Trees and Shrubs Hardy in the Cooler Temperate Regions of the Northern Hemisphere**. Jamaica Plain, Mass.: Arnold Arboretum of Harvard University, 1949; repr., Monticello, N.Y.: Lubrecht and Cramer, 1978. 825p. $84.00. ISBN 3874291286. (Collectanea Bibliographica Series, No. 9).

The purpose of this bibliography is to give reference to sources of botanical names, valid names, and synonyms of the woody plants. It is a companion volume to the author's *Manual of Cultivated Trees and Shrubs* (1940), now out-of-print and superseded by Little (entry 514) or Sargent (entry 541).

116. Saccardo, Pier A. **Sylloge Fungorum omnium hucusque cognitorum**. New York: Johnson, 1972. 25v. $1,950.00/set; $1,800.00/set(pa.). ISBN 0384528317; 0384528309pa. (Vol. 26: 1972. $210.00. ISBN 0685136124).

This reprint of the 1931 edition is a catalog of names with Latin descriptions of fungi. It predates Petrak's *List* (entry 114) and aims to list all the genera and species of fungi published up to 1920. This work is updated by *Saccardo's Omissions*, edited by Paul M. Kirk (Slough, England: Commonwealth Agricultural Bureaux, 1985, 101p., $34.00).

117. Stafleu, Frans A., and Richard S. Cowan. **Taxonomic Literature: A Selective Guide to Botanical Publications and Collections with Dates, Commentaries and Types**. 2nd ed. Utrecht, Netherlands: Bohn, Scheltema and Holkema, 1976- . indexes. (Regnum Vegetabile, Vols. 94, 98, 105, 110, 112-). (Vol. 5: $163.50).

This is one of the most important guides to the taxonomic literature and it contains a wealth of other pertinent information. Arranged alphabetically by author, data includes brief biographical information, herbaria where collections are held, location of bibliography and biographical entries, composite works, eponymy, location of handwriting specimens, and annotated list of authors' important publications with complete bibliographical details. There are name and title indexes for each volume. This is a highly specialized reference work, one that is indispensable for the taxonomist/detective trying to locate elusive coauthors, titles, collections, and writings of significant botanists of the past. A core collection of 5,000 of the rarer and most useful nonjournal titles from this set is available on microfiche from Meckler Publishing (Westport, Conn., 1985-86, $42,500.00/set). It is also available on a subscription basis from Meckler. [R: ARBA, 1984, entry 1304]

118. **Taxon**. See entry 193.

119. **Taxonomic Index. 1939-1956**. Vols. 1-19. Lancaster, Pa.: American Society of Plant Taxonomists.

This index to botanical taxonomic literature was continued in *Brittonia* from 1957-67 (see entry 151). The index was arranged by author within systematic categories.

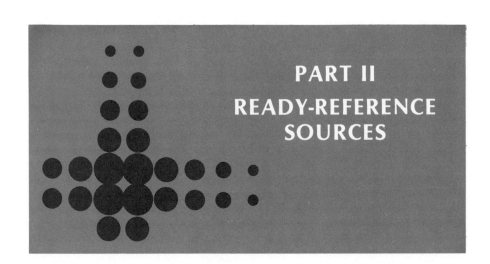

PART II
READY-REFERENCE
SOURCES

3 Current
Awareness Sources

This chapter is divided into sections for different types of current awareness sources: periodicals or journals; sources of book reviews, patents, translations; and reviews or yearbooks.

Journals

Periodicals constitute a primary, extremely important, resource in the literature of the botanical sciences. Serial literature, and in particular the journal literature, reports results of original research, creating a dependence on periodical publications that cannot be overlooked or overemphasized in terms of impact and influence. Both the descriptive and functional botanical literature use the periodical information transfer route to disseminate primary research results which stand as a matter of record, making the refereed journal the standard in terms of priority rights, whether it be a new species or a new synthesis of photosynthesis biochemistry that is in question.

Historically, botany was one of the two most important, and comparatively independent, of the biological sciences; it was the basic science for agriculture (Brown, 1956). In an informative essay on the nature of botanical serials, Brown summarized his findings: (1) botany is dependent on journals in general science, agriculture, chemistry, and medicine; (2) the botanical literature is stable and uses a larger percentage of books to journals than do other scientific disciplines; (3) chemistry and physics are both assuming more importance in the study of botany. Although Brown's book was published in 1956, much of what he found is still true at the present time. It is a fact of scientific life that botanists, as a group, use research from literally thousands of different journals; in botany, as many "classic" papers are published in multidisciplinary journals as are published in the designated "botany" journals (Garfield, 1977). A list of important journals from other disciplines heavily used by botanists includes the following periodicals from general science, biochemistry, ecology, and microbiology.

120. **American Naturalist**. Vol. 1- . Chicago: University of Chicago Press, 1867- . monthly. $90.00/yr. ISSN 0003-0147.

121. **Biochemical Journal**. Vol. 1- . Colchester, England: Biochemical Society, 1911- . 2 pts. twice per month. $1,010.00/yr. pts. 1 and 2. Pt. 1, **Cellular Aspects**: ISSN 0306-3283. Pt. 2, **Molecular Aspects**: ISSN 0306-3275.

122. **Biochemistry**. Vol. 1- . Washington, D.C.: American Chemical Society, 1964- . fortnightly. $303.00/yr. ISSN 0006-2960.

123. **Biochimica et Biophysica Acta**. Vol. 1- . Amsterdam: Elsevier Scientific Publishing, 1947- . irregular. 42v./yr. $2,925.51/yr. all sections. ISSN 0006-3002.

124. **Ecology**. Vol. 1- . Tempe, Ariz.: Ecological Society of America, 1920- . bimonthly. $85.00/yr. ISSN 0012-9658.

125. **Evolution**. Vol. 1- . Lawrence, Kans.: Society for the Study of Evolution, 1947- . bimonthly. $100.00/yr. ISSN 0014-3820.

126. **Experimental Cell Research**. Vol. 1- . New York: Academic, 1950- . 14/yr. $792.00/yr. ISSN 0014-4827.

127. **Journal of Bacteriology**. Vol. 1- . Washington, D.C.: American Society for Microbiology, 1916- . monthly. $280.00/yr. ISSN 0021-9193.

128. **Journal of Biological Chemistry**. Vol. 1- . Bethesda, Md.: American Society of Biological Chemists, 1905- . twice per month. $360.00/yr. ISSN 0021-9258.

129. **Journal of Cell Biology**. Vol. 1- . New York: Rockefeller University Press, 1955- . monthly. $250.00/yr. ISSN 0021-9525.

130. **Journal of Ecology**. Vol. 1- . Oxford, England: Blackwell, 1913- . triannual. $155.00/yr. ISSN 0022-0477.

131. **Nature**. Vol. 1- . New York: Macmillan, 1869- . weekly. $240.00/yr. ISSN 0028-0836.

132. **Proceedings of the National Academy of Sciences**. Vol. 1- . Washington, D.C.: National Academy of Sciences, 1915- . twice per month. $215.00/yr. ISSN 0027-8424.

133. **Science**. Vol. 1- . Washington, D.C.: American Association for the Advance ment of Science, 1880- . weekly. $98.00/yr. ISSN 0036-8075.

134. **Virology**. Vol. 1- . New York: Academic, 1955- . 16/yr. $680.00/yr. ISSN 0042-6822.

In terms of historical significance, *Curtis's Botanical Magazine* is the oldest botanical journal in continuous publication. It was founded in 1787 in London by William Curtis, the owner and editor, who was an eminent English botanist and entomologist. At that time it was called *The Botanical Magazine or Flower-Garden Displayed*; "in which the most ornamental foreign plants, cultivated in the open ground, the greenhouse, and the stove, are accurately represented in their natural colours. To which will be added, their names, class, order, generic and specific characters, according to the celebrated Linnaeus; their places of growth, and times of flowering: together with the most approved method of culture. A work intended for the use of such ladies, gentlemen, and gardeners, as wish to become scientifically acquainted with the plants they cultivate." Beginning with volume 15 the magazine was sold, and hence-forth was known as *Curtis's Botanical Magazine*, employing as editors, some of the foremost English botanists of the day.

Currently, *Curtis's* is incorporated into *The Kew Magazine*, begun in 1984, and published by the Royal Botanic Gardens, Kew, in association with Collingridge Books, Feltham, Middlesex, England. Besides the features so successfully included in *Curtis's* for so many years, *Kew Magazine* has added other, special attractions for botantists, horticulturists, plant ecologists, and botanical illustrators. It is published quarterly, at $45.00/yr. for U.S. subscribers. *Curtis's* has been known from its inception for its truly outstanding botanical color illustrations. *Kew Magazine* intends to continue this tradition.

Although this chapter will not dwell on important botanical journals of the past, several journals should be mentioned in passing as dominating the nineteenth- and early twentieth-century botanical literature. All of these early volumes should be considered obligatory for botanical research collections. They are available on microfiche from Inter Documentation Company AG, Switzerland.

> *Botanische Jahrbucher für Systematik, Pflanzengeschichte und Pflanzengeographie.* Leipzig, Germany: 1881-1930.
>
> *Botanische Zeitung.* Leipzig, Berlin: 1843-1910.
>
> *Botaniska Notiser.* Lund, Sweden: 1839-1930.
>
> *Bulletin of the Torrey Botanical Club.* New York: 1870-1930.
>
> *Flora, oder allgemeine botanische Zeitung.* Regensburg, Germany: 1818-1930. Preceded by *Botanische Zeitung.*
>
> *Gardeners' Chronicle.* London: 1841-73. New York: 1874-86.
>
> *Journal of Botany, British and Foreign.* London: 1863-1930.

The remainder of the journal section lists and briefly annotates the most significant currently published journals of botany.

135. **Acta Biologica Cracoviensia. Botanica**. Vol. 1- . Warsaw, Poland: Publishing House of the Polish Academy of Sciences, 1958. 2/yr. price varies. ISSN 0001-5296.
Original papers presented before the Academy are published in English.

136. **Acta botanica neerlandica**. Vol. 1- . Leiden, Netherlands: Royal Botanical Society of the Netherlands, 1952. bimonthly. $50.00/yr. ISSN 0044-5983.
Text mainly in English. This is a continuation of the *Nederlandsch kruidkundig archief* (1846-1951) and the *Receuil des travaux botaniques neerlandais* (1904-51). The journal contains original articles in all areas of botany, brief communications, abstracts of papers presented at the meetings of the Royal Botanical Society of the Netherlands, book reviews, and announcements.

137. **Acta Oecologica-Oecologia Plantarum**. Vol. 5- . Montreuil Cedex, France: Gauthier-Villars, 1970- . quarterly. $142.88/yr. for three series (*Oecologia Applicata*; *Oecologica Generalis*; *Oecologica Plantarum*). ISSN 0243-7678.
Formerly (until 1980) *Oecologica Plantarum*. Text in English, French or German; summaries in English and French.

138. **American Journal of Botany**. Vol. 1- . Miami, Fla.: Botanical Society of America, 1914- . 10/yr., May-June and Nov.-Dec. issues combined. $65.00/yr. ISSN 0002-9122.
This journal, "devoted to all branches of plant sciences," publishes manuscripts in three areas: articles that are reports of original research; special papers, including reviews, evaluations on critical subjects, and recent advances in a specialized research area; and rapid communications, short papers reporting research of timely interest. In 1983 and 1984, the May-June issues contained the program, with abstracts of papers, of the annual meeting of the Botanical Society of America.

139. **Angewandte Botanik**. Vol. 1- . Berlin: Paul Parey, 1919- . 3 double issues per year. DM 298,00/yr. ISSN 0066-1759.
Text in German; summaries in English. This journal publishes original articles, book reviews, personal notes, and a subject index in each issue.

140. **Annals of Botany**. Vol. 1- . New York: Academic, 1887. 12/yr. $246.00/yr. ISSN 0305-7364.
Contributions on any aspect of plant science are considered for publication; both experimental and theoretical papers are welcome. Book reviews are included.

141. **Annals of the Missouri Botanical Garden**. Vol. 1- . St. Louis, Mo.: Missouri Botanical Garden, 1914- . quarterly. $70.00/yr. ISSN 0026-6493.
This journal publishes original papers, primarily in systematic botany, from the Missouri Botanical Garden, although outside articles will be considered. Potential authors should contact the *Annals* editor for more information; page charges are assessed.

142. **Australian Journal of Botany**. Vol. 1- . Melbourne, Australia: Commonwealth Scientific and Industrial Research Organisation, 1953- . bimonthly. $90.00/yr., with supplements. ISSN 0067-1924.
Original papers in any branch of botany relevant to the Australian region are acceptable. Descriptive articles and critical review papers are considered for the

Journal, while more lengthy manuscripts, for example, major taxonomic revisions, are normally published in the supplementary series.

143. **Australian Journal of Plant Physiology**. Vol. 1- . Melbourne, Australia: Commonwealth Scientific and Industrial Research Organisation, 1974- . bimonthly. $80.00/yr. ISSN 0310-7841.

Results of original research are published in "all the main areas of plant physiology ranging from biochemistry and biophysics to whole-plant and environmental physiology and including structural, genetical and other aspects as related to function."

144. **Berichte der deutschen botanischen Gesellschaft**. Vol. 1- . Stuttgart, Germany: Fischer, 1882- . 3/yr. $110.00/yr. ISSN 0365-9631.

Text and summaries in German and English.

145. **Biochemie und Physiologie der Pflanzen (BPP)**. Vol. 1- . Jena, Germany: Fischer, 1970- . 9/yr. $143.00/yr. ISSN 0015-3796.

Text in English, French, and German; summaries in English. Formerly *Flora*, this journal is published as the A section of *Flora*: physiology and biochemistry (see entry 158). It will accept manuscripts in related fields of plant biology such as microbial biochemistry and physiology, cell physiology, molecular biology, genetics, biophysics, and functional topography. "Regular" papers are welcome, as well as short communications and reviews.

146. **Botanical Bulletin of Academia Sinica (Taipei)**. Vol. 1- . Nankang, Taiwan: Academia Sinica, Institute of Botany, 1960- . 2/yr. $14.00/yr. ISSN 0006-8063.

Text in English; summaries in Chinese. Original contributions may be from any field of botany.

147. **Botanical Gazette**. Vol. 1- . Chicago: University of Chicago Press. quarterly. $70.00/yr. (institutions). ISSN 0006-8071.

Original research in the plant sciences is published reporting experimental work; techniques or methods are not considered unless accompanied by illustrative observations or data. Evaluative papers concerned with new perspectives of plant biology are welcome.

148. **Botanical Journal of the Linnean Society**. Vol. 1- . New York: Academic, 1855- . 8/yr. $260.00/yr. ISSN 0024-4074.

Published for the Linnean Society of London. The *Journal* "publishes papers of relevance to, and reviews of, the taxonomy of all plant groups, including anatomy, biosystematics, cytology, ecology, electron microscopy, morphogenesis, palaeobotany, palynology and phytochemistry."

149. **Botanical Magazine, Tokyo**. Vol. 1- . Tokyo: Botanical Society of Japan, 1887- . quarterly. $100.00/yr. ISSN 0006-808X.

Text in English. This international journal considers original manuscripts reporting research in all areas of plant science.

150. **Botanical Review**. Vol. 1- . New York: New York Botanical Garden, 1935- . quarterly. $40.00/yr. ISSN 0006-8101.

Authors are chosen primarily by invitation, but unsolicited manuscripts are also considered. This journal bills itself as "interpreting botanical progress" and serves the function of synthesizing the state of knowledge of the botanical sciences for a broad spectrum of botanists.

151. **Brittonia.** Vol. 1- . New York: New York Botanical Garden, 1931- . quarterly. $40.00/yr. ISSN 0007-196X.

This journal of systematic botany accepts research papers on systematic botany in the broadest sense and includes relevant fields such as anatomy, botanical history, chemotaxonomy, cytology, ecology, morphology, paleobotany, palynology, phylogenetic systematics, and phytogeography. News, announcements, and book reviews are an integral part of the publication.

152. **Bryologist.** Vol. 1- . Buffalo, N.Y.: American Bryological and Lichenological Society, 1898- . quarterly. $50.00/yr. (institutions). ISSN 0007-2745.

With text in English, French, German, and Spanish and summaries in English, this publication is "devoted to the study of bryophytes and lichens." Full or partial page costs are assessed depending on author's financial support. Checklists for nomenclature of mosses appear in this journal periodically.

153. **Bulletin of the Torrey Botanical Club.** Vol. 1- . New York: New York Botanical Garden, 1870- . quarterly. $35.00/yr. ISSN 0040-9618.

Each issue contains three parts: part 1 includes original research papers; part 2 called *Torreya*, includes general, invited, review papers, and papers on local flora, field trip reports, obituaries, book reviews, notes, and short papers on conservation and environmental concerns; and part 3 is the "Index to American Botanical Literature" compiled from the resources of the New York Botanical Garden Library. The index is arranged alphabetically by author within broad subject areas, such as algae, spermatophytes, economic botany, and paleobotany.

154. **Canadian Journal of Botany/Journal Canadien de botanique.** Vol. 1- . Ottawa: National Research Council of Canada, 1929- . monthly. $150.00/yr. ISSN 0008-4026.

Text in English or French. Research papers or notes and a limited number of invited review articles on topics of current interest are accepted. All areas of botany are included.

155. **Canadian Journal of Plant Science/Revue Canadien de phytotechnie.** Vol. 1- . Ottawa: Agricultural Institute of Canada, 1921- . quarterly. $42.42/yr. ISSN 0008-4220.

The official journal of the Canadian Society of Agronomy and the Canadian Society for Horticultural Science. Original research results on any aspects of plant science will be considered. Manuscripts will be accepted in English and in French.

156. **Economic Botany.** Vol. 1- . New York: New York Botanical Garden, 1947- . quarterly. $40.00/yr. ISSN 0013-0001.

Published for the Society for Economic Botany. "Devoted to past, present, and future uses of plants by man," with emphasis on scientific papers relating to "uses" rather than growing of plant materials. Articles dealing with agronomy or horticulture are not included.

157. **Environmental and Experimental Botany**. Vol. 1- . Elmsford, N.Y.: Pergamon Press, 1961- . quarterly. $155.00/yr. ISSN 0098-8472.

The journal publishes papers from all fields of botany in (1) experimental areas: radiation botany, photobotany, chemical mutagenesis, cell proliferation, anatomy and morphology, cytogenetics, somatic cell genetics, and statistical and mathematical procedures and models; and (2) environmental areas: pollution effects, plant-soil/water/atmosphere/temperature relations, phytopathology, gravitational botany, species dispersal and propagation. Original research papers, review articles, software information, and symposium proceedings are included in the publication.

158. **Flora: Morphologie, Geobotanik, Oekophysiologie**. Vol. 1- . Jena, Germany: Fischer, 1818- . bimonthly. $100.00/yr. ISSN 0367-2530.

Text in English, French, German; summaries in English and German. Original articles on plant structure (morphology and anatomy) and on plant distribution and plant communities (geobotany) are considered for publication. Taxonomic papers will be accepted if their content is related to morphology, geobotany, or ecology. Experimental ecological articles are especially welcome if they deal with survival in particular habitats. Purely systematic and nomenclatural articles or problems of local interest will not be accepted. *Flora*, the oldest German scientific botanical journal, formerly consisted of two sections: (A) physiology and biochemistry, and (B) morphology and geobotany. Since 1970, part A has been published under the title *Biochemie und Physiologie der Pflanzen* (*BPP*) (see entry 145). Section B from volume 159 is published under *Flora*.

159. **History and Philosophy of the Life Sciences**. Section II of **Pubblicazioni della statzione zoologica di Napoli**. Vol. 1- . Italy: Olschki, 1979- . semiannual. $36.00. ISSN 0391-9714.

This international journal is devoted to the historical development and social and epistemological implications of the life sciences; its scope also includes the philosophical concerns of biology and medicine.

160. **Israel Journal of Botany**. Vol. 1- . Jerusalem: Weizmann Science Press, 1951- . quarterly. $45.00/yr. ISSN 0021-213X.

Text in English. Although this journal is international in scope in all areas of the plant sciences, special emphasis is given to topics related to the Mediterranean and Near Eastern regions. Some issues contain book reviews and abstracts of papers presented at related scientific meetings.

161. **Journal of Bryology**. Vol. 7- . Oxford, England: Blackwell, 1972- . semiannual. $68.50/yr. ISSN 0373-6687.

Formerly, *Transactions of the British Bryological Society*, published for the Society. This journal reports original research concerning all facets of the study of mosses and liverworts. Articles may take the form of research papers or the shorter "Bryological Notes." There is a book review section and each issue devotes space to a listing of the recent bryological literature compiled from the Society's library, *Biological Abstracts*, *The Bryologist*, *Cryptogamie: Bryologie, Lichenologie,* and *Referativnyi Zhurnal*.

162. **Journal of Experimental Botany**. Vol. 1- . Oxford, England: Oxford University Press, 1950- . monthly. $230.00/yr. ISSN 0022-0957.

Published for the Society for Experimental Botany and designated as an official journal of the Federation of European Societies for Plant Physiology. The journal acts as a medium for the publication of original papers and occasional review articles in the fields of plant physiology, biochemistry, biophysics, and related topics. Book reviews form a part of each issue.

163. **Journal of Natural Products (Lloydia).** Vol. 1- . Cincinnati, Ohio: Lloyd Library and Museum and the American Society of Pharmacognosy, 1938- . bimonthly. $90.00/yr. ISSN 0163-3864.

Volumes 1-41 were titled *Lloydia*. All aspects of natural products research are welcome, including research that relates to biochemistry and the biology of living systems. Papers are accepted in three formats: full papers, notes, or brief communication. Manuscripts are accepted in French or German, although English is preferable. Book reviews and meeting announcements are included in each issue.

164. **Journal of Phycology.** Vol. 1- . Lawrence, Kans.: Allen, 1965- . quarterly. $95.00/yr. ISSN 0022-3646.

Text in English, French, German, or Spanish; summaries in English. Published for the Phycological Society of America. The *Journal* publishes original research articles relating to the taxonomy, ecology, morphology, cytology, physiology, and biochemistry of algae. Review articles are by invitation only; preliminary notes, progress papers, and methods or techniques articles are not acceptable. An annual supplement records abstracts of papers from the annual meeting of the Phycological Society of America.

165. **Journal of Plant Growth Regulation.** Vol. 1- . New York: Springer-Verlag, 1981- . quarterly. $80.00/yr. ISSN 0721-7595.

This journal deals with the control of growth and development from cell to plant community by natural hormones and synthetic regulators, in both fundamental and applied aspects. For an informative positive review of this journal in comparison to its competitors, see "Limits to Growth" by L. J. Audus (*Nature*, 305, no. 5934, Oct. 7, 1983, p. 489).

166. **Journal of Plant Physiology.** Vol. 115- . Stuttgart, Germany: Fischer, 1984- . 3/yr. DM 246,00/vol. ISSN 0176-1617.

Formerly *Zeitschrift für Pflanzenphysiologie*, 1909-83. This international journal of plant physiology publishes original articles, short communications, and reviews on all aspects of plant physiology. English is preferred, but manuscripts in German and French are accepted. Applied papers are accepted on the condition that they contribute toward the understanding of basic physiological problems.

167. **Journal of the Arnold Arboretum.** Vol. 1- . Cambridge, Mass.: Arnold Arboretum, Harvard University, 1919- . quarterly. $50.00/yr. ISSN 0004-2625.

Staff papers have priority in this publication, although papers are accepted from former students, staff, or botanists who have worked at the Arboretum, or who have reported work on a plant group or geographic area of interest to the Arboretum.

168. **Mycologia.** Vol. 1- . New York: New York Botanical Garden, 1909- . bimonthly. $60.00/yr. ISSN 0027-5514.

Published for the Mycological Society of America. Authors are restricted to those who have been a member in good standing of the Society for over one year immediately preceding submission of the manuscript, although exceptions have been made upon a vote by the Editorial Board. Original research, notes or brief articles reporting research or new techniques, and invited papers dealing with fungi and lichens are considered. Book reviews and obituaries are included in the journal.

169. **Mycotaxon**. Vol. 1- . Ithaca, N.Y.: Mycotaxon, Ltd., 1974- . quarterly. $21.00/ yr. individual; $48.00/yr. institution. ISSN 0093-4666.

"An international journal designed to expedite publication of research on taxonomy and nomenclature of fungi and lichens." Papers may be in French or English with summaries in any language.

170. **New Phytologist**. Vol. 1- . New York: Academic, 1902- . monthly. $361.80/yr. ISSN 0028-646X.

Articles may report research on any subject related to botany and must be in English. Long or review articles are considered upon consultation with the editor. Book reviews form a section of the journal.

171. **Nordic Journal of Botany**. Vol. 1- . Copenhagen: Council for Nordic Publications in Botany, 1981- . bimonthly. $77.95/yr. ISSN 0107-055X.

Formed by the merger of *Botanisk Tidsskrift* (1866-1981), *Botaniska Notiser* (1839-1981), *Friesia* (1932-81), and *Norwegian Journal of Botany* (1952-81). Text in English. This journal, representing botanists in Denmark, Finland, Norway, and Sweden, is published in cooperation with *Oikos, Ornis Scandinavica, Holarctic Ecology, Lindbergia, Physiologia Plantarum*, and the monograph series *Opera Botanica*. Papers are acceptable in the following areas of botany: holarctic and general taxonomy, tropical taxonomy, geobotany, structural botany, mycology, lichenology, and phycology. According to a favorable review by P. D. Moore (*Nature*, 299, no. 5883, Oct. 7, 1982, p. 508), there is strong emphasis on taxonomic subjects with very little ecological research included.

172. **Photosynthetica**. Vol. 1- . Prague, Czechoslovakia: Czechoslovak Academy of Sciences, 1967- . quarterly. $96.00/yr. ISSN 0300-3604.

Text in English (preferentially), French, German; summaries in English. An international journal for photosynthesis research. Original papers in all fields of photosynthesis are accepted, as well as brief communications, reviews on specialized topics, book reviews, bibliographies of reviews, reports of conferences and meetings, and methodological papers.

173. **Physiologia Plantarum**. Vol. 1- . Copenhagen: Munksgaard, 1948- . monthly. $246.00/yr. ISSN 0031-9317.

Text in English, French, or German. Official publication of the Scandinavian Society for Plant Physiology and of the Federation of European Societies of Plant Physiology. The journal publishes papers on all aspects of plant physiology, including biochemistry, biophysics, classical plant physiology, and ecology. Mini-reviews on photosynthesis-energy metabolism, uptake-translocation, growth, and development should concentrate on the state-of-the-art and/or the personal view of the author rather than providing complete coverage of the literature. Prospective authors for the mini-review series should contact one of the editors before submission of the manuscript. Beginning in December 1984, "What's New in Plant Physiology," brief descriptions of timely topics of interest to plant physiologists, will become part of *Physiologia Plantarum*.

174. **Physiological and Molecular Plant Pathology.** Vol. 28- . New York: Academic, 1986- . bimonthly. $220.00/yr. ISSN 0885-5765.

Formerly *Physiological Plant Pathology* from volumes 1-27 (1971-85), this international journal of experimental plant pathology publishes research on "physiological, biochemical and molecular aspects of host-parasite interactions, and environmental, genetical and ultrastructural studies that help to elucidate these processes." Papers on any kind of infective agent are acceptable, but they must have a direct bearing on host-parasite interaction. Manuscripts breaking new conceptual ground or clearly advancing the understanding of the molecular control of the host-parasite interaction are especially welcome. Only papers in English are accepted; summaries may be in the language of the author's choice.

175. **Physiologie Vegetale.** Vol. 1- . Montreuil Cedex, France: Gauthier-Villars, 1963- . bimonthly. $102.00/yr. ISSN 0031-9368.

Text in English, French, or German; summaries in English and French. Published for the Société Française de Physiologie Végétale. Original contributions are accepted on theoretical, experimental, and technical papers dealing with the physiology, biochemistry, structure, and genetics of plant physiology at the molecular, subcellular, cellular, organ, whole plant, or environmental level of integration. The following are welcome: original articles, reviews, letters to the editor, technical articles, and brief reports of papers presented at meetings of the Société. To be retitled *Plant Physiology and Biochemistry* in 1987.

176. **Phytochemistry.** Vol. 1- . Elmsford, N.Y.: Pergamon Press, 1962- . monthly. $420.00/yr. ISSN 0031-9422.

Text in English, French, and German. An international journal of plant biochemistry, it is an official organ of the Phytochemical Society of Europe. Society announcements appear at regular intervals. This journal publishes research on all aspects of pure and applied plant biochemistry, especially those factors underlying the growth, development, and differentiation of plants and the chemistry of plant products. Contributions must be in English and are accepted as full papers, short reports, or reviews. Prospective authors of reviews should consult with the editors before submission of the manuscript. The journal is divided into five sections: biochemistry, biosynthesis, chemotaxonomy, phytochemistry, and short reports. There is also a section for book reviews.

177. **Phyton: Annales Rei Botanicae.** Vol. 1- . Horn, Austria: F. Berger, 1948- . annual. DM 110/yr. ISSN 0079-2047.

Scientific papers in German, English, or French from all fields of the plant sciences are considered. Reviews are acceptable with the editor's consent.

178. **Phytopathologische Zeitschrift/Journal of Phytopathology.** Vol. 1- . Berlin: Paul Parey, 1930- . monthly. $429.50/yr. ISSN 0031-9481.

Text in English, German, French, and Italian with summaries in English and German. Contents include original papers, short communications, and book reviews covering all aspects of phytopathology and associated fields.

179. **Phytopathology**. Vol. 1- . St. Paul, Minn.: American Phytopathological Society, 1911- . monthly. $150.00/yr. ISSN 0031-949X.

The official journal of the American Phytopathological Society. Membership in the Society is not a prerequisite for publication, but mandatory page charges are higher for nonmembers than for members. Papers are accepted in the following categories: (1) original research on some aspect of plant pathology, accounts of techniques, and phytopathological history of general importance; (2) review papers concerning new concepts, hypothesis, theory, or other integration of plant pathology; and (3) letters to the editor explaining, amplifying, or commenting upon research published in the journal. The journal also provides reports of the officers, meetings, and news of the Society.

180. **Plant and Cell Physiology**. Vol. 1- . Kyoto, Japan: Japanese Society of Plant Physiologists, 1960- . 8/yr. $110.00/yr. ISSN 0032-0781.

Text in English. Contributors must hold membership in the Japanese Society of Plant Physiologists. This international journal publishes original papers pertaining to physiology, biochemistry, molecular biology, and cell biology of plants and microorganisms.

181. **Plant and Soil**. Vol. 1- . The Hague: Nijhoff/Junk. 18/yr. $472.00/yr. ISSN 0032-079X.

Text in English, French, and German. Issued under the auspices of the Royal Netherlands Society of Agricultural Science. "An international journal of plant nutrition, plant-microbe associations, soil microbiology, and soil-borne plant diseases." Book reviews are not included.

182. **Plant Cell Reports**. Vol. 1- . New York: Springer-Verlag, 1981- . bimonthly. $96.00/yr. ISSN 0721-7714.

This journal presents original research results in biochemistry, physiology, molecular biology, cytology, genetics and genetic engineering, phytopathology, and morphogenesis. Contributions are published on the average within six weeks of receipt of the final version.

183. **Plant Disease**. Vol. 64- . St. Paul, Minn.: American Phytopathological Society, 1980- . monthly. $110.00/yr. ISSN 0191-2917.

"An international journal of applied plant pathology," this official journal of the American Phytopathological Society, continues *Plant Disease Reporter*. Although this · journal is in the applied area and is important in the study of agriculture, it must be considered one of the key sources for the botanical sciences as a whole.

184. **Plant Molecular Biology**. Vol. 1- . The Hague: Nijhoff/Junk, 1981- . monthly. $135.00/yr. ISSN 0167-4412.

An international journal on fundamental research and genetic engineering published in cooperation with the International Society for Plant Molecular Biology, this journal provides a "rapid publication outlet for all types of research concerned and connected with plant molecular biology and plant molecular genetics." Papers should report studies on higher and lower plants, including cyanobacteria, algae, fungi, and yeast. Both fundamental and applied plant research is welcome. Developments in animal molecular biology, bacterial molecular biology, and other related fields will form a regular section of short reviews. A section on plant biotechnology news and views will deal with techniques and advances in plant genetic engineering, with the addition of patents, nitrogen fixation, and plant tissue culture information, as warranted. Short essays discussing the impact of biotechnology also will be included in the journal.

185. **Plant Pathology**. Vol. 1- . Oxford, England: Blackwell, 1952- . quarterly. $130.00/yr. ISSN 0032-0862.

An international journal edited by the British Society for Plant Pathology. Research papers and critical reviews in all areas of plant pathology are accepted; short reports of new or unusual records of plant diseases are also published. Book reviews form a regular section of the journal.

186. **Plant Physiology**. Vol. 1- . Rockville, Md.: American Society of Plant Physiologists, 1926- . monthly. $330.00/yr. ISSN 0032-0889.

Full-length papers and short communications reporting research in all phases of plant physiology are considered for publication. One contributor from each manuscript must hold membership in the Society. Authors are charged a handling fee of $100.00 for each accepted manuscript.

187. **Plant Science**. Vol. 38- . Limerick, Ireland: Elsevier Scientific Publishing, 1985- . monthly. $655.00/yr. ISSN 0168-9452.

Formerly *Plant Science Letters*, 1973-84. Original work in any area of experimental plant biology is welcome. Areas covered include cell and tissue culture, cell organelles, cellular and molecular genetics, enzymology, inorganic metabolism, molecular biology, nucleic acids, photosynthesis, phytochemistry, pigments, regulation of growth and development, ultrastructure, viruses, and water relations. "An international journal of experimental plant biology."

188. **Plant Systematics and Evolution/Entwicklungsgeschichte und Systematik der Pflanzen**. Vol. 1- . New York: Springer-Verlag, 1851- . monthly. $300.00/yr. ISSN 0378-2697.

A continuation of *Österreichische botanische Zeitschrift*. Text in English and German. Original papers on the morphology and systematics of plants are accepted, with preference given to papers dealing with "modern methods of cytogenetics, population analysis, electron microscopy, chemosystematics, palynology, and numerical analysis." The editors emphasize that a comparative, developmental, or functional approach is essential.

189. **Planta**. Vol. 1- . New York: Springer-Verlag, 1925- . monthly. $796.00/yr. ISSN 0032-0935.

Text in English, French, and German, with summaries in English. This international journal publishes original articles in structural and functional botany, plant physiology, and population botany. Phytopathological and genetics papers are accepted contingent upon their contribution toward botanical problems.

190. **Planta Medica**. Vol. 1- . Stuttgart, Germany: Thieme, 1953- . bimonthly. $145.00/yr. ISSN 0032-0943.

This journal of medicinal plant research publishing original articles and reports is the official organ of the Society for Medicinal Plant Research. This periodical contains the Society's "Newsletter," published three times per year, since February 1984. Contributions to the "Newsletter" are welcome and will be accepted in English, German, or French.

191. **Soviet Plant Physiology**. Vol. 1- . New York: Consultants Bureau, Plenum, 1962- . monthly. $725.00/yr. ISSN 0038-5719.

English translation of *Fiziologiya Rastenii* from the Academy of Sciences of the U.S.S.R. The journal appears about six months after the publication of the original Russian issue and carries the same number and date as the original Russian issue.

192. **Systematic Botany**. Vol. 1- . Kent, Ohio: American Society of Plant Taxonomists, 1976- . quarterly. $60.00/yr. ISSN 0363-6445.

This journal accepts theoretical and applied papers on taxonomic botany and related areas from members of the Society. An editorial fee may be assessed for long manuscripts and authors are encouraged to contribute toward the cost of publishing. Announcements and book reviews are included in each issue of the journal.

193. **Taxon**. Vol. 1- . Utrecht, Netherlands: International Bureau for Plant Taxonomy and Nomenclature, 1951- . quarterly. $86.00/yr. ISSN 0040-0262.

Published for the International Association for Plant Taxonomy, the journal is "devoted to systematic and evolutionary biology with emphasis on botany." Issues contain original articles, proposals to conserve or reject nomenclature and to revise the International Code of Botanical Nomenclature, chromosome number reports, Flora Neotropica news, current pteridological research, book reviews and announcements, news and notes.

194. **Transactions of the British Mycological Society**. Vol. 1- . Cambridge, England: Cambridge University Press, 1896- . monthly. $375.00/yr. ISSN 0007-1536.

Transactions publishes original research articles in all areas of mycology, notes and brief articles, as well as book reviews.

195. **Vegetatio**. Vol. 1- . The Hague: Kluwer/Junk, 1949- . monthly. $309.00/yr. ISSN 0042-3106.

Text in English, French, German, and Spanish. The journal is the official organ of the International Society for Vegetation Science and publishes articles in the field of geobotany, vegetation science, and plant population ecology. Book reviews, notices and summaries of proceedings of scientific meetings, are included.

Sources of Book Reviews, Patents, Translations

Selection of materials has always been an intriguing problem. This chapter annotates several general sources for acquisition and book reviews with the aim of listing these tools more for the benefit of students and the general public than for librarians. There is no attempt to consider the process and progress of library acquisition in any degree of depth. Chapter 1 discusses those bibliographic tools designed especially for botany; the library tools annotated in this chapter are of general use to a wide range of botanists and librarians for selection and verification of materials both scientific and popular.

Although all of the sources discussed in this chapter may be considered to be alerting devices to materials new to the searcher, only patents are considered *primary* literature, that is, literature that conveys new information. Recently, with the increased emphasis on genetic engineering and recombinant DNA techniques, patent and translations of primary literature have assumed critical importance; sources in this chapter have been chosen to aid in their retrieval.

Verification

196. Aslib Book List: A Monthly List of Recommended Scientific and Technical Books, With Annotations. Vol. 1- . London: Aslib, 1935- . quarterly. $65.00/yr. ISSN 0001-2521.

Critical book reviews.

197. Associations' Publications in Print. Vol. 1- . Ann Arbor, Mich.: Bowker, 1981- . annual. $137.50/yr. ISSN 0000-0663.

This series provides information on print and nonprint materials published by over four thousand United States and Canadian associations. There are subject, publisher, and title indexes to information on pamphlets, journals, newsletters, bulletins, books, and other printed materials published by national, state, regional, local, and trade associations. It is also available online as APIP, a computerized database allowing information to be retrieved by title, subject, publisher, association, status, frequency, country, format, special features, and circulation.

198. Books in Print. New York: Bowker, 1948- . annual. $170.00/yr. ISSN 0068-0214.

This basic verification tool includes complete bibliographic information for in-print and forthcoming books from U.S. publishers. It is current, comprehensive, and updated annually, sometimes more frequently, depending on the particular publishing schedule. It includes all subjects and levels, with books listed by author and title. Companion sets, *Subject Guide to Books in Print*, *Forthcoming Books*, and *Scientific and Technical Books and Series in Print*, for example, are available from the same publisher. A directory for all U.S. publishers is included in each set. A computerized database, BOWKER BOOKS IN PRINT (BBIP), is available as the online version of the printed tools. BBIP provides bibliographic information for books in print from 1979 to date, for forthcoming books within the next six months, for books out-of-print or out-of-stock, and is updated monthly or more frequently during peak publishing periods.

199. **Books in Series**. New York: Bowker, 1985. 6v. $325.00/set. ISBN 0835219380.

The first two editions were published under the title *Books in Series in the United States*. "Original, reprinted, in-print, and out-of-print books, published or distributed in the U.S. in popular, scholarly, and professional series." This does not include series for children, primary or secondary school texts, or U.S. government publications. There are series, author, and title indexes.

200. **Bowker's Microcomputer Software in Print**. New York: Bowker, 1985. 2v. $95.00. ISBN 0835219445.

This guide to microcomputer programs is the *Books in Print* for microcomputer software. It lists 20,000 programs under 100 major applications and some 2,500 specific subcategories with full descriptive information for software pertaining to education, the home, business, and industry published by major suppliers as well as independents. Also see *The Software Catalog: Science and Engineering* (entry 385).

201. **Cumulative Book Index: A World List of Books in the English Language**. Vol. 1- . New York: H. W. Wilson, 1928- . monthly, except August, with annual cumulations. service basis. ISSN 0011-300X.

This publication includes all English-language imprints of the year designated, except government publications, maps, sheet music, pamphlets, and other ephemera. Complete bibliographic information is provided, which is invaluable and indispensable for verification.

202. **Directory of Published Proceedings, Series SEMT-Science/Engineering/Medicine/Technology**. Vol. 1- . White Plains, N.Y.: InterDok, 1965- . monthly. $325.00/yr. ISSN 0012-3293.

Included are preprints and published proceedings of congresses, conferences, symposia, meetings, seminars, and summer schools held since 1964. Arrangement is chronological with subject and sponsor indexes; complete bibliographic information is provided.

203. **Dissertation Abstracts International. B: The Sciences and Engineering**. Vol. 1- . Ann Arbor, Mich.: University Microfilms International, 1969- . monthly. $170.00/yr. ISSN 0419-4217.

Abstracts of dissertations on microfilm from cooperating, academic, mostly U.S. institutions are included. Key word, title, and author indexes are provided. *A Comprehensive Dissertation Index*, 1861-1972, is available from the same publisher. DISSERTATION ABSTRACTS ONLINE (DAI), the computerized version of the printed database, provides abstracts and complete access to bibliographic information for dissertations from 1861 to date.

204. **Index to Scientific and Technical Proceedings**.

For complete bibliographic information, see entry 87. Unlike *Directory of Published Proceedings*, this publication indexes at the individual article level by author, author's organization, conference sponsor, subject, meeting location, and title words. It is a unique resource.

205. **National Union Catalog: U.S. Books**. Washington, D.C.: Library of Congress Catalog Management and Publication Division, 1983- . monthly. $245.00/yr. ISSN 0734-7642.

This is "a repertory of the cataloged holdings of selected portions of the cataloged collections of the major research libraries of the United States and Canada, plus the more rarely held items in the collections of selected smaller and specialized libraries." Indispensable for verification and holdings information.

206. **National Union Catalog, 1956 through 1967: A Cumulative Author List Representing Library of Congress Printed Cards and Titles Reported by Other American Libraries**. Totowa, N.J.: Rowman and Littlefield, 1970-72. 125v. $2,750.00/set. ISBN 0874710006.

"A new and augmented twelve-year catalog being a compilation into one alphabet of the fourth and fifth supplements of the *National Union Catalog* with a key to additional locations through 1967 and with a unique identifying number allocated to each title."

207. **New Serial Titles: A Union List of Serials Commencing Publication after December 31, 1949**. Washington, D.C.: Library of Congress, 1953- . irregular. $350.00/yr.

Serials are arranged by title with complete bibliographic and holding information from U.S. cooperating libraries. This is a basic source that updates the third edition of the *Union List of Serials*.

208. **Ulrich's International Periodicals Directory: A Classified Guide to Current Periodicals, Foreign and Domestic**. New York: Bowker, 1932- . annual. $110.00/yr. ISSN 0000-0175.

This is available in the annual printed version and as a computerized database (ULRICH'S), which is updated every six weeks. The database profiles over 117,000 regularly and irregularly issued serial publications from 65,000 publishers in 181 countries. Information includes all titles connected to a particular periodical over its span of publication, dates, frequency, publisher, circulation, and indexing services.

209. **Union List of Serials in Libraries of the United States and Canada**. 3rd ed. Edited by Edna Brown Titus. New York: H. W. Wilson, 1965. 5v. $175.00/set.

This is an important source for serial verification and holdings information and for interlibrary loan. It is updated by *New Serial Titles*. For convenience, it is optimal to have the complete serial title in hand; serials published by a corporate body are entered under location.

210. **World List of Scientific Periodicals Published in the Years 1900-1960**. 4th ed. Edited by P. Brown and G. B. Stratten. Washington, D.C.: Butterworths, 1963-65. 3v. Vol. 1: 531p. $139.30. ISBN 0317419196. Vol. 2: 1186p. $160.00. ISBN 031741920X. Vol. 3: 1824p. $160.00. ISBN 0317419218.

This is a union list of periodicals held by British libraries. Complete bibliographic information is included with the accepted abbreviations. As the titles are arranged by abbreviation, they can be accessed by a truncated version of a less-than-completely accurate citation. The *World List* is easy to use, authoritative, comprehensive, and of great assistance with "mystery" citations. Supplements are available.

Book Reviews

211. **Book Review Digest: An Index to Reviews of Current Books**. Vol. 1- . New York: H. W. Wilson, 1905- . monthly, except February and July. service basis. ISSN 0006-7326.

This periodical provides abstracts of book reviews for current fiction and nonfiction in the English language. Although it excludes technical scientific books, the *Digest* does include books on science for the general reader. Online retrieval is available through WILSONLINE.

212. **Book Review Index**. Vol. 1- . Detroit: Gale, 1965- . bimonthly. $160.00/yr. ISSN 0524-0581.

This index scans 445 publications to cover book reviews in the social sciences, humanities, natural sciences, fine arts, arts and crafts, business and economics, religion, and philosophy. It is appropriate for reviews of interest for the general reader. A machine-readable database, BRI, is available from DIALOG.

213. **Choice**. Vol. 1- . Chicago: Association of College and Research Libraries, 1964- . 11/yr. $100.00/yr. ISSN 0009-4978.

This is an excellent book reviewing periodicals appropriate for college- and university-level materials. Coverage is broad; reviews are classified by subject and arranged alphabetically thereafter. Reviews are indexed in *Book Review Index*, *Book Review Digest*, and *Library Literature*.

214. **Science Books and Films**. Vol. 1- . Washington, D.C.: American Association for the Advancement of Science, 1965- . 5/yr. $20.00/yr. ISSN 0098-342X.

Also called *AAAS Science Books and Films*, this publication reviews trade books, textbooks, 16-mm films, videocassettes, and filmstrips for elementary, junior and senior high school and college students, general audiences, and professionals. It covers the sciences and mathematics for all ages, except textbooks for kindergarten through senior high school. Evaluative reviews rate each item as highly recommended, recommended, acceptable, or not recommended with the appropriate audience level indicated. There is a separate section for the botanical sciences. This is a comprehensive source for reliable reviews of scientific books and films.

215. **Technical Book Review Index**. Vol. 1- . Pittsburgh, Pa.: JAAD Publishing, 1935- . monthly, September through June. $43.00/yr. ISSN 0040-0890.

This index identifies book reviews in current scientific, technological, medical, and trade journals, and provides quotes from the reviews. It covers pure science, life science, medicine, agriculture, and technology. This index compares favorably with *Science Books and Films* (see entry 214), *ARBA* (see entry 2), and *Choice* (see entry 213) as the most valuable, evaluative sources for technical and scientific books for the student or professional botanist.

Patents

In addition to the specific sources listed below for domestic and international patent information, consult these computerized databases: CAB ABSTRACTS, LIFE SCIENCES COLLECTION, and NTIS.

216. **Biological Abstracts and Biological Abstracts/RRM.**
For complete bibliographic information and annotation, see entries 69 and 70. Beginning in 1986, U.S. patent information is provided in the printed and online version of the BIOSIS PREVIEWS database.

217. **CASSIS (Classification and Search Support Information System).** Washington, D.C.: U.S. Department of Commerce, Patent and Trademark Office, 1984- .
This free online patent information system provides Patent Depository libraries direct, online access to Patent and Trademark Office data, including original and cross-reference classifications of a patent, all patents assigned to a classification, classification titles, search of key words in classification titles, patent abstracts, and classification index terms. CASSIS is designed to assist in using patent collections in Patent Depository libraries and this service is available from these libraries.

218. **Chemical Abstracts.** Vol. 1- . Columbus, Ohio: Chemical Abstracts Service, 1907- .
See entry 76 for a complete annotation. *Chemical Abstracts* is an excellent source for U.S. and foreign patent information. It includes patent assignee, granting country, application country, date of issue, patent number, classification number, and patent application number in the country of priority application.

219. **INPADOC.** McLean, Va.: Pergamon Infoline. price available upon request.
This computerized database covers equivalent patents in foreign countries from 1968 to the present including over ten million patent records. Access is provided by free text, International Patent Classification subject code, assignee, or inventor.

220. **Official Gazette of the United States Patent and Trademark Office.** Vol. 1- . Washington, D.C.: Government Printing Office, 1872- . weekly. $375.00/yr. ISSN 0098-1133. **Supplement**: $238.00/yr. ISSN 0360-5132.
This weekly listing of patents issued includes a brief description and sketch for each patent. There are patentee, classification, and geographical indexes. The annual *Index to Patents Issued from the United States Patent and Trademark Office* lists patentees, reissue patentees, design patentees, plant patentees, defensive publications and disclaimers, and dedications.

221. **PATSEARCH.** McLean, Va.: Pergamon Infoline. price available upon request.
A computerized database of U.S. Patents and Patent Cooperation Treaty (PCT) published applications covering U.S. utility patents since 1970, reissue patents since 1975, design patents since 1976, defensive publications since 1976, and all PCT published applications and search reports. Other online patent files from Pergamon include INPADOC, complete bibliographic information on world patent documents since 1968; INPANEW, new patent documents; COMPUTERPAT, a companion to PATSEARCH, covering all data processing patents assigned to selected U.S.

subclasses; and PATLAW, headnotes to intellectual property case law and administrative decisions reported in the *U.S. Patents Quarterly* since 1967.

222. **PLANT PATENT.** Vol. 1- . Washington, D.C.: U.S. Patent Office, 1931- . irregular.

This publication consists of plant patents granted by the United States. It lists inventor, assignee, application number, filing date, abstract, discovery information, description of plant and flower, and provides color photographs for each plant. It is arranged by plant number and can be accessed through the *Index to Patents*. Individual patent descriptions can be purchased through private vendors or the Patent Trademark Office.

223. **WORLD PATENTS INDEX.** London: Derwent Publications. price available upon request.

This machine-readable database, available from DIALOG, is based on the printed publications *Central Patents Index*, *WPI Gazette Service*, *World Patents Abstracts*, and *Electrical Patents Index*. Patent documents from foreign countries and the United States are covered in two files for chemical, general, electrical, and mechanical subject areas. Depending on the subject, patent information from 1963 to date is included.

Translations

224. **Journals in Translation.** 2nd ed. Wetherby, England: British Library, Lending Division, 1978. 181p. index. ISBN 0853501718.

Bibliographic information and annotations are listed for 982 journals that are available in a translated version. There are appendixes for Soviet patents, the *Doklady*, and the Institute of Electrical Engineers in Japan.

225. **Transdex Index.** Vol. 1- . Wooster, Ohio: Bell & Howell, 1974- . monthly or annual microfilm cumulation. $645.00/yr. ISSN 0041-1116.

This publication indexes translations published as part of the Joint Publications Research Service of the U.S. government.

226. **Translations Register-Index.** Vol. 1- . Chicago: National Translations Center, The John Crerar Library of the University of Chicago, 1967- . monthly. $115.00/yr. ISSN 0041-1256.

This periodical lists new translations added to the depository of the John Crerar Library, which acts as a repository and information clearinghouse for more than 500,000 translations into English of scientific and technical literature from 40 different languages. *Consolidated Index of Translations into English II* cumulates the years 1967-84 and is available on microfiche (24X) from the Center for $300.00 or as three softbound printed volumes for $450.00.

Reviews

Scientific review articles, monographs, or books are very different in scope and content from the book reviews discussed earlier in this chapter which basically discuss or evaluate a particular book's usefulness, quality, scope, etc. Review articles may be defined as a "subject survey of the primary literature usually covering the material that was published since the most recent review of the same subject." (Bonn, 1982).

Articles or books that review a particular subject form a very important part of the botanical literature. Review articles can be used to introduce a subject to the novice or to experts who are investigating a topic peripheral to their own research. A review article can correlate the literature and provide a summary of an area to offer a starting point for future work in the field. It can provide a useful bibliography or set of references for access to classic work conducted earlier. Reviews, which can be great time savers, can critique, evaluate, or they can simply present information, allowing readers to form their own opinion. Access to review articles or edited books is gained through the use of abstracting and indexing serial publications, such as *Bibliography of Agriculture*, *Biological and Agricultural Index*, *Biological Abstracts*, *Chemical Abstracts*, and *Index to Scientific Reviews*, discussed more completely in chapter 2.

Following are brief annotations for a selected list of the most significant scientific reviews for the botanical sciences.

227. **Advances in Botanical Research**. Vol. 1- . New York: Academic, 1963- . irregular. $65.00/yr. ISSN 0065-2296.

This well-established series traditionally surveys progress across the whole spectrum of botanical studies. However, the editors are considering a more thematic nature for issues after volume 11 (1985).

228. **Advances in Bryology**. Vol. 1- . Vaduz, Liechtenstein: J. Cramer, 1981- . biennial. (Vol. 1: 1981. $60.00. ISBN 3768212963).

The official publication of the International Association of Bryologists. This series aims to present authoritative and current reviews, essays, and summary syntheses of the different fields of bryology written by leaders in the area. Most articles provide extensive references to the literature.

229. **Advances in Economic Botany**. Vol. 1- . Bronx, N.Y.: New York Botanical Garden, 1984- . irregular. ISSN 0741-8280. (Vol. 1: 1984. $28.00. ISBN 0893272531).

This series was established to provide an outlet for monographs and symposia on all subjects in the field of economic botany. The subjects covered are of interest to all levels of workers. Although this series is of peripheral interest to basic botanical science, it is included here because it does present a blend of essential information and practical application. Volume 1 is titled *Ethnobotany in the Neotropics*.

230. **Advances in Photosynthesis Research: Proceedings of the Sixth International Congress on Photosynthesis, Brussels, Belgium, August 1-6, 1983**. Edited by C. Sybesma. The Hague: Nijhoff/Junk; distr., Hingham, Mass.: Kluwer Academic, 1984. 4v. ill. bibliog. indexes. $478.00/set. ISBN 9024729467/set. (Advances in Agricultural Biotechnology).

Thirty chapters provide an overview of state-of-the-art work in photosynthesis research, including fundamental theory and applied aspects. Most of the chapters give information concerning recent developments in the field, although there are some contributions from the associated Round Table Discussion on Light-Controlled Development of the Photosynthetic Apparatus held in Antwerp, July 29-30, 1983.

231. **Annual Review of Phytopathology**. Vol. 1- . Palo Alto, Calif.: Annual Reviews, 1963- . annual. $31.00/yr. ISSN 0066-4286.

This is a highly respected series devoted to review articles by individual authors, dealing with plant pests and diseases, their history, nature, effects, ecology, and control. Most articles have extensive bibliographies to provide broad access to the field. [R: ARBA, 1981, entry 1422]

232. **Annual Review of Plant Physiology**. Vol. 1- . Palo Alto, Calif.: Annual Reviews, 1950- . annual. $31.00/yr. ISSN 0066-4294.

This series is addressed to the advanced student doing research in plant physiology and plant biochemistry. There are subject and author indexes for each volume and cumulated indexes produced for the previous five volumes. All articles have extensive literature references. [R: ARBA, 1981, entry 1421]

233. **Bioessays**. Vol. 1- . Cambridge, England: Cambridge University Press, 1984- . monthly. $140.00/yr. ISSN 0265-9247.

This is a review journal for molecular, cellular, and developmental biologists. The October 1985 issue is devoted to plant molecular biology.

234. **Biotechnology Advances**. Vol. 1- . New York: Pergamon Press, 1983- . semi-annual. $110.00/yr. ISSN 0734-9750.

This series provides research reviews and patent abstracts for bionics and genetic engineering.

235. **Botanical Monographs**. Vol. 1- . Oxford, England: Blackwell, 1961- . irregular. price varies.

Each volume provides a specific review for a particular topic, for example, volume 23: *Plant Cell Culture Technology*, 1985.

236. **Botanical Review**. Vol. 1- . Bronx, N.Y.: New York Botanical Garden, 1935- . quarterly. $40.00/yr. ISSN 0006-8101.

This review journal interprets the state of knowledge in the botanical sciences. Most articles are by invitation.

237. **CRC Critical Reviews in Plant Sciences**. Vol. 1- . Boca Raton, Fla.: CRC Press, 1983- . quarterly. $104.00/yr. ISSN 0735-2689.

This review series focuses on providing critical reviews on photosynthesis, nitrogen fixation, genetic engineering, and other areas of plant science. It covers basic as well as applied state-of-the-art surveys.

238. **Current Concepts in Plant Taxonomy**. Edited by Vernon H. Heywood and D. M. Moore. Orlando, Fla: Published for the Systematics Association by Academic Press, 1984. 432p. ill. bibliog. index. $50.00. ISBN 0123470609. (Systematics Association Special Volume, No. 25, ISSN 0309-2593).

This volume is based on papers presented at an International Conference on Current Concepts in Plant Taxonomy, organized under the auspices of the Systematics

Association and held at the University of Reading on July 7-9, 1982. It presents a contemporary overview of challenges, problems, achievements, and advances in special fields of plant taxonomy by leading experts in the field. [R: QRB, Dec. 1985, p. 507]

239. **Current Topics in Medical Mycology.** Vol. 1- . New York: Springer-Verlag, 1985- . annual. $79.50/vol. (Vol. 1: 1985. ISBN 0387960953).

Topics of current interest to medical mycologists are summarized in this review series. Each volume can serve as a survey of contemporary advances, future research directions, and uses of medically important fungi in basic and applied science.

240. **Economic and Medicinal Plant Research.** Vol. 1- . Edited by H. Wagner, Hiroshi Hikino, and Norman R. Farnsworth. Orlando, Fla.: Academic, 1985- . (Vol. 1: 295p. $72.50. ISBN 0127300600).

The intent of volume 1 of this edited treatise is "to identify areas of research in natural plant products that are of immediate or projected importance from a practical point of view. . . ." The reviews are concise and critical. [R: QRB, Sept. 1986, p. 411]

241. **Encyclopedia of Plant Physiology, New Series.** Vol. 1- . New York: Springer-Verlag, 1975- . irregular. price varies.

This important monographic review series surveys all aspects of the botanical sciences, devoting several volumes, as needed, to cover any particular topic. The new series is in English, continuing the older German series, *Handbuch der Pflanzenphysiologie*, volumes 1-18, 1955-67. Recent volumes in this excellent series include:

Vol. 9: *Hormonal Regulation of Development I: Molecular Aspects of Plant Hormones.* 1980. 691p. $148.00. ISBN 0387101616.

Vol. 10: *Hormonal Regulation of Development II: The Functions of Hormones from the Level of the Cell to the Whole Plant.* 1984. 309p. $76.00. ISBN 0387101969.

Vol. 11: *Hormonal Regulation of Development III: Role of Environmental Factors.* 1985. approx. 950p. $139.50. ISBN 0387101977.

Vol. 12, A-D: *Physiological Plant Ecology I-IV.* 1981-83. (A: $120.00. ISBN 0387107630. B: $133.00. ISBN 0387109064. C: $125.00. ISBN 0387109072. D: $130.00. ISBN 0387109080).

Vol. 13, A-B: *Plant Carbohydrates I-II.* 1981-82. (A: $145.00. ISBN 0387110607. B: $127.00. ISBN 0387110070).

Vol. 14, A-B: *Nucleic Acids and Proteins in Plants I-II.* 1982. (A: $130.00. ISBN 0387110089. B: $130.00. ISBN 0387111409).

Vol. 15: *Inorganic Plant Nutrition.* 1983. 2v. $150.00/set. ISBN 038712103X.

Vol. 16: *Photomorphogenesis.* 1983. 2v. $150.00/set. ISBN 0387121439.

Vol. 17: *Cellular Interactions.* 1984. 743p. $140.00. ISBN 0387127380.

Vol. 18: *Higher Plant Cell Respiration.* 1985. approx. 550p. $104.50. ISBN 0387139354.

242. Foyer, Christine H. **Photosynthesis.** New York: John Wiley, 1984. 219p. ill. index. $29.95. ISBN 0471864730. (Cell Biology, Vol. 1).

This comprehensive review covers the breadth of important topics during photosynthesis. [R: QRB, Sept. 1985, p. 356]

243. Grant, Verne. **Genetics of Flowering Plants**. New York: Columbia University Press, 1975. 514p. ill. bibliog. $52.50; $22.00pa. ISBN 0231036949; 0231083637pa.

In the author's words, this is a "summary of our knowledge and understanding of the genetics of higher plants." Emphasis is placed on the nature and action of genes, gene systems, linkage systems, and genetic systems in higher plants and includes a presentation of a series of classic genetic experiments in higher plants beginning with Mendel.

244. **Handbook of Vegetation Science**. Edited by Reinhold Tuxen. Continued by H. Lieth. Vol. 1- . The Hague: Junk, 1974- . irregular. price varies. (Vol. 3: 666p. ISBN 9061931843).

This multivolume reference work reviews vegetation science in all its many aspects. Topics range from history, philosophy, methods, structure and function of communities to practical applications.

245. **Handbuch der Pflanzenanatomie: Encyclopedia of Plant Anatomy: Traite d' anatomie vegetale**. 2nd ed. Berlin: Gebruder-Borntraeger, 1934- . irregular. price varies.

Some of the more recent volumes in this prestigious, scholarly series include:

Vol. 8, Pt. 2B, Pt. 1: *Anatomie des blattes. II Blattanatomie der angiospermen*. 1984. ISBN 3443140149.

Vol. 10, Pt. 1: *Fruits of Angiosperms*. 1977. ISBN 3443140106.

Vol. 13, Pt. 1: *Anatomie des galles*. 1983. ISBN 3443140130.

246. Jones, Evan Benjamin Gareth. **Recent Advances in Aquatic Mycology**. New York: John Wiley, 1976. 748p. ill. bibliog. index. $99.95. ISBN 0470291761.

This review of the aquatic fungi stands the test of time. It is favorably reviewed in *Nature* (May 6, 1976, p. 82) and *Science* (Nov. 19, 1976, p. 833).

247. **Modern Methods in Plant Taxonomy**. Edited by Vernon H. Heywood. New York: Published for the Botanical Society of the British Isles by Academic Press, 1968. 312p. $59.50. ISBN 0123469503. (Botanical Society of the British Isles Conference Report, No. 10).

This major survey of plant taxonomy, circa 1968, is based on papers delivered at the Conference of the Botanical Society of the British Isles. Emphasis is placed on techniques of study.

248. **Mycotoxic Fungi, Mycotoxins, Mycotoxicoses: An Encyclopedic Handbook**. Edited by Thomas D. Wyllie and Lawrence G. Morehouse. New York: Marcel Dekker, 1977-78. 3v. ill. bibliog. indexes. Vol. 1: $99.75. ISBN 0824765508. Vol. 2: $99.75. ISBN 0824765516. Vol. 3: $69.75. ISBN 0824765524.

This three-volume set provides an introduction to, and a review of, the complex field of mycotoxicology. It is international in scope, broad in coverage, and includes chapters by experts from eleven countries. [R: QRB, June 1979, pp. 189-90]

249. **New Manual of Bryology**. Edited by Rudolf M. Schuster. Nichinan, Japan: Hattori Botanical Laboratory, 1983. 2v. ill. bibliog. index. Vol. 1: 11,500Y. Vol. 2: 12,500Y. ISBN 49381633045/set.

The present status of the field of bryology is evaluated by specialists in this cooperative effort. For more complete treatment of physiological ecology of bryophytes, see volume 1 of *Advances in Bryology* (entry 228).

250. **Oxford Surveys of Plant Molecular and Cell Biology**. Vol. 1- . New York: Oxford University Press, 1984- . annual. $45.00/yr. ISSN 0264-861X.

This series contains review articles delineating current progress, news, and views in the field of plant molecular and cell biology.

251. **Plant Biochemistry**. Edited by Donald Henry Northcote. London: Butterworths, 1974. 287p. ill. index. ISBN 0839110758. (Biochemistry Series 1, Vol. 11, MTP International Review of Science).

This excellent review covers aspects of the literature of plant biochemistry through 1973. For more current treatment, see *Progress in Phytochemistry* (entry 254), *Recent Advances in Phytochemistry* (entry 255), *Annual Proceedings of the Phytochemical Society of Europe* (entry 438), and plant biochemical texts in chapter 9. [R: QRB, Sept. 1976, p. 443]

252. **Progress in Botany**. Vol. 36- . New York: Springer-Verlag, 1974- . irregular. price varies. ISSN 0340-4773.

Continues *Fortschritte der Botanik*, Vol. 1-35, 1931-73. The basic mission of this series is to report on all areas of botany, rotating subjects every two to three years. [R: QRB, June 1976, pp. 319-20]

253. **Progress in Phycological Research**. Vols. 1-2. Amsterdam: Elsevier Scientific Publishing, 1982-83. Vol. 3- . Bristol, England: Biopress, Ltd., 1984- . irregular. price varies. ISSN 0167-8574.

This series provides in-depth surveys of algae research and will interest taxonomic, ecological, or cell phycologists.

254. **Progress in Phytochemistry**. Vols. 1-3. New York: Wiley Interscience, 1968-72. Edited by Leonora Reinhold, et al. Vol. 4- . New York: Pergamon Press, 1977- . irregular. price varies. ISSN 0079-6689. (Vol. 7: 1981. $96.00).

The purpose of this series is to provide a clear, critical, integrated evaluation of the recent advances in phytochemistry, including sufficient background information for the nonspecialist.

255. **Recent Advances in Phytochemistry: Proceedings of the Phytochemical Society of North America**. Vols. 1-4. New York: Appleton-Century-Crofts, 1968-72. Vols. 5-8. New York: Academic Press, 1972-74. Vol. 9- . New York: Plenum, 1975- . irregular. price varies. ISSN 0079-9920. (Vol. 19: **Chemically Mediated Interactions between Plants and Other Organisms: Proceedings of the 24th Annual Symposium**. 1985. 246p. $45.00. ISBN 0306420066).

Each volume is based on the proceedings of the annual symposium and reviews a particular topic in depth.

256. **Studies in Biology**. Vol. 1- . London: Published for the Institute of Biology by Edward Arnold, 1966- . irregular. $8.95/vol. ISSN 0537-9024.

This well-respected review series aims to advance both the science and practice of biology. In addition to other activities, the Institute publishes primary journals, conducts examinations, arranges meetings, represents views of professional bodies to the government and other associations in the United Kingdom. *Studies in Biology* reports significant developments, selected lists of books for further reading, and suggestions for practical work. Each book is well written by an expert in the field and is concise, reviewing concepts in an interesting, informative way for the student and the teacher. An example is *Genetic Engineering in Higher Plants* by J. Roger Warr (1984, 58p., ISBN 0713128852pa., Institute of Biology's Studies in Biology, No. 162).

257. **Topics in Photosynthesis**. Vol. 1- . Amsterdam: Elsevier Scientific Publishing, 1976- . annual. price varies. ISSN 0378-6099.

This series was instituted to help specialists and nonspecialists alike keep abreast of progress and development in particular fields of photosynthesis. Each volume has a distinctive title: volume 6 (1985) is titled *Photosynthetic Mechanisms and the Environment*.

References

Bonn, George S. 1982. Literature of science and technology. In *McGraw-Hill encyclopedia of science and technology*, 5th ed., Vol. 7, 754-60. New York: McGraw-Hill.

Brown, Charles Harvey. 1956. *Scientific serials: Characteristics and lists of most-cited publications in mathematics, physics, chemistry, geology, physiology, botany, zoology, and entomology*. ACRL Monograph 16. Chicago: Association of College and Research Libraries.

Garfield, Eugene. 1977. *Essays of an information scientist*. Vol. 2, 1974-76. Philadelphia: Institute for Scientific Information.

4 Dictionaries and Encyclopedias

This list of dictionaries and encyclopedias is broadly construed to contain several entries that are equivalent to each other. For example, if Longman's is not available, Chinery is a reasonable alternative. For some questions it may be necessary to consult several sources; for example, for the satisfactory resolution to a problem in Russian terminology, this dictionary list provides several choices for the difficult or ephemeral word. Whenever possible, the preferred source will be designated and/or the important differences between choices will be identified.

258. **Ainsworth and Bisby's Dictionary of the Fungi**. 7th ed. By D. L. Hawksworth, B. C. Sutton, and G. C. Ainsworth. Kew, England: Commonwealth Mycological Institute; distr., Forestburgh, N.Y.: Lubrecht and Cramer, 1983. 445p. ill. $27.50. ISBN 0851985157.

This is an essential dictionary for anyone working with fungi or lichens. There are approximately 16,500 entries comprising the most complete listing of generic names of fungi and lichens that is currently available as of the publication date. Information includes the date of publication, status, systematics, number of species, distribution, and reference to key publications. Identification patterns of families, orders, and higher categories are included for most groups.

259. Alex, Jack F., et al. **Common and Botanical Names of Weeds in Canada/Noms populaires et scientifiques des plantes nuisibles du Canada.** rev. ed. Ottawa: Research Branch, Agriculture Canada, 1980. 132p. ISBN 0660505479. (Publication/Agriculture Canada, No. 1397).

This volume is written in both English and French and includes common and botanical names of weeds native to Canada. "Weeds" have been interpreted broadly to include weeds in the usual sense as well as many herbaceous and woody plants of range, pasture, forest, and water. Code numbers have been assigned to each variety of weed by the Expert Committee on Weeds for use in their computerized retrieval system.

260. Baranov, A. **Basic Latin for Plant Taxonomists.** Lehre, Germany: J. Cramer, 1958; repr., Monticello, N.Y.: Lubrecht and Cramer, 1971. 146p. $16.00pa. ISBN 3768207277.

Although this is not as useful as Stearn (see entry 304), it does serve to instruct the plant taxonomist in preparing formal, Latin descriptions for new taxa. The guide is written from a practical point of view with emphasis on methods for writing plant descriptions, especially descriptions of species, and is intended for beginners in plant taxonomy.

261. Bedevian, Armenag K. **Illustrated Polyglottic Dictionary of Plant Names in Latin, Arabic, Armenian, English, French, German, Italian and Turkish Languages, Including Economic, Medicinal, Poisonous and Ornamental Plants and Common Weeds.** Cairo: Argus & Papazian, 1936. 1099p.

Part 1 of this two-part volume includes 3,657 entries in English arranged by scientific name, followed by abbreviated name of the author, synonyms, family name, and common name in seven languages. Part 2 contains indexes of the common names of plants, referring to entry number in the first part of the dictionary. This is a useful source with an unusual selection of languages. Unfortunately, it may be out-of-print with no complete substitute immediately at hand.

262. Borror, Donald J. **Dictionary of Word Roots and Combining Forms, Compiled from the Greek, Latin, and Other Languages, with Special Reference to Biological Terms and Scientific Names.** Palo Alto, Calif.: Mayfield, 1960. 134p. $5.95pa. ISBN 0874840538.

This dictionary is designed to "meet the needs of the beginning student, the medical student, and the taxonomist." The dictionary is arranged alphabetically with concise information for root derivation, word endings, and meaning. There are short sections for formulation of scientific names, translation of Greek words, and some common combining forms. This dictionary is of value for taxonomic botanists, although Jaeger's *Source-Book* is superior (see entry 286).

263. Cash, Edith K. **A Mycological English-Latin Glossary.** New York: Hafner, 1965. 152p. (Mycologia Memoir No. 1, published for the New York Botanical Garden in collaboration with the Mycological Society of America).

This book is written to aid in preparing Latin diagnoses of fungi. Gender and genitive case of nouns, endings of adjectives, principal parts of verbs, and the cases governed by propositions are included. This unusual dictionary is useful in the special case described when only an elementary knowledge of Latin grammar is required. For more complete assistance, see Stearn (entry 304).

264. Chinery, Michael. **A Science Dictionary of the Plant World: An Illustrated Demonstration of Terms Used in Plant Biology**. New York: F. Watts, 1969. 264p. ill. (col.).

Over 400 color pictures illustrate this dictionary useful for the beginning student.

265. Coon, Nelson. **Dictionary of Useful Plants**. Emmaus, Pa.: Rodale Press, 1974. 290p. ill. bibliog. ISBN 087857090X.

Arranged alphabetically under plant families, this dictionary provides information on cultivated and wild plants native to the United States. Description, habitat, uses, and customs are included for each plant. [R: ARBA, 1976, entry 1373]

266. Crosby, Marshall R., and Robert E. Magill. **Dictionary of Mosses: An Alphabetical Listing of Genera Indicating Familial Disposition, Nomenclatural and Taxonomic Synonymy Together with a Systematic Arrangement of the Families of Mosses and a Catalogue of Family Names Used for Mosses**. 2nd ed. St. Louis, Mo.: Missouri Botanical Garden, 1978. 43p. (Monographs in Systematic Botany from the Missouri Botanical Garden, Vol. 3).

This dictionary provides a means of identifying quickly the family in which any genus of mosses is placed. Synonyms, orthographic variants, and names for genera accepted as taxonomic entities are included; only names which have been validly published are included. It updates V. F. Brotherus's treatment published in 1924-25 ("Musci," in *Die natürlichen Pflanzenfamilien* by A. Engler and K. Prantl, Leipzig, Germany) and *Index Muscorum* (entry 105). This short dictionary also includes a systematic arrangement of the families of mosses, an index to their systematic arrangement, a catalogue of family names used for mosses, and a brief list of references.

267. Cunningham, John J., and Rosalie J. Coté. **Common Plants: Botanical and Colloquial Nomenclature**. New York: Garland, 1977. 120p. ISBN 0824099060. (Reference Library of Science and Technology Series, Vol. 3).

This very interesting dictionary is divided into two sections, the botanical nomenclature of common plants and the origins of plant names. There is a glossary of selected species epithets, an index of botanical nomenclature listing the common name with the botanical names, colloquial names of plants, commemorative names, and a lengthy section on the history and folklore of the common names of selected plants. This book links the world of the horticulturalist and layperson with the botanist; it is useful for both of these areas, but it suffers in comparison with Howes (see entry 284).

268. Davydov, Nikolai N. **Botanical Dictionary: Russian-English-German-French-Latin**. 2nd ed. Moscow: n.p., 1962; repr., Forestburgh, N.Y.: Lubrecht and Cramer, 1981. 335p. $28.00. ISBN 3874291979.

Thirty percent of the 6,000 entries in Russian are the names of plants with references given to the families to which they belong. All entries are numbered and each language has an index.

269. Debenham, Clifford N. **The Language of Botany**. Sydney, Australia: Society for Growing Australian Plants, c1962, 1971. 208p. ISBN 0909830010.

Another of the general botanical dictionaries suitable for the student or layperson. Illustrations are sparse, but the terms may fill in gaps left by other dictionaries.

270. Durant, Mary. **Who Named the Daisy? Who Named the Rose? A Roving Dictionary of North American Wildflowers**. New York: Congdon & Weed, 1983. 224p. ill. $8.95pa. ISBN 0312929447.

Although this is not a comprehensive guide, it does provide interesting information, name origins, and pleasing drawings for more than 100 common North American wildflowers. [R: ARBA, 1978, entry 1289]

271. Edlin, Herbert L., Maurice Nimmo, and Eric Bourdo. **The Illustrated Encyclopedia of Trees, Timbers and Forests of the World**. New York: Harmony, 1978. 256p. ISBN 0517534509.

There are discussions of trees and timbers, guides to conifers, broadleaves, and tropical and southern hemisphere trees. The color photographs and illustrations are beautifully done; there is a short glossary and an index. This is an alternative for the *Oxford Encyclopedia of Trees* (entry 294) because their similarities outweigh their differences.

272. **Elsevier's Dictionary of Botany**. Compiled by Paul Macura. New York: Elsevier Scientific Publishing, 1979-82. 2v. (Vol. 1: $117.00. ISBN 0444417877).

Volume 1 (1979) lists plant names in English, French, German, Latin, and Russian, while volume 2 (1982) lists general terms in English, French, German, and Russian. This multilingual dictionary contains almost 16,000 entries from botany and related fields of agricultural chemistry, phytobiology, phytochemistry, horticulture, and taxonomy; it covers plants, trees, shrubs, mushrooms, lichens and other, more general terms. The numbered entries are arranged alphabetically in English; indexes for the other languages refer to the numbered items. Each volume is designed for use alone and cross-references for synonyms are included. The aim is to provide a broad, comprehensive dictionary suitable for translation of scholarly work and this dictionary is an invaluable reference work.

273. **Elsevier's Lexicon of Plant Pests and Diseases: Latin, English, French, Italian, Spanish, German**. Compiled and arranged by M. Merino-Rodriguez. Amsterdam: Elsevier Scientific Publishing, 1966. 351p.

The purpose of this dictionary is to provide a comprehensive and systematic list of current usage for plant pests and diseases. The lexicon is taxonomically arranged in two sections: zooparasites of plants and phytoparasites of plants. There are three appendixes for symptoms of disease, nonparasitic disease, and unclassified virus disease. Six language indexes refer the reader to the main section of the dictionary.

274. **Encyclopaedia Botanica: The Definitive Guide to Indoor Blossoming and Foliage Plants**. Compiled by Dennis A. Brown. New York: Dial Press, 1978. 304p. ISBN 0385270402.

Besides basic information on botany, this encyclopedia discusses plant culture, seasonal care, propagation, pests, disease, herbs, and plants for special uses. Both the common and scientific names are provided for each plant and there is a section devoted to current plant societies. There is a glossary and an index. This is a useful introduction for the novice; it is included here because of its usefulness in locating names and information on indoor plants. For comprehensive treatment of exotic plants for the home or patio, consult Graf (entry 281). For more emphasis on herbs, see either Kadans (entry 287) or Morton (entry 292).

275. **Encyclopedia of Herbs and Herbalism**. Edited by S. Malcolm. New York: Grossett and Dunlap, 1979. 304p. ISBN 0448154722.

This outstanding book discusses cultivation, preservation, uses, history, folklore, and chemistry of herbs and includes superb color photographs, drawings, and diagrams. The treatment is authoritative and it is highly recommended by several reviewers. [R: ARBA, 1981, entry 1441; BL, June 1980, p. 1565]

Although this edition is out-of-print, an abridged edition is available, *Van Nostrand Reinhold Color Dictionary of Herbs and Herbalism* (New York: Van Nostrand Reinhold, 1982, 160p., $14.95, ISBN 0442283385).

276. **Encyclopedia of Mushrooms**. Edited by Colin Dickinson and John Lucas. New York: Putnam, 1979. 280p. ISBN 0399121048.

There are introductory chapters on the history, biology, life-style, habitat, identification, edible properties, and utilization of mushrooms. The 148-page reference section includes colored illustrations, discussions of the characteristics of the fruiting body, habitat and distribution, occurrence, culinary properties, scientific and common names, and indexes. It is written with emphasis on natural history and is appropriate for the nonspecialist. [R: ARBA, 1980, entry 1390]

277. **Encyclopedia of the Plant Kingdom**. Edited by Anthony Huxley. New York: Chartwell, 1977. 240p. ISBN 0890090874.

This encyclopedia is arranged by subject under two sections: (1) the living world of plants and (2) some important plant groups. The book was written by experts for the general reader and provides excellent color photographs to illustrate the reading material. A glossary, index, and brief bibliography are included.

278. **Facts on File Dictionary of Botany**. Edited by Stephen Blackmore and Elizabeth Tootill. New York: Facts on File, 1984. 400p. $21.95. ISBN 0871968614.

Over 3,000 entries are arranged alphabetically with relatively few illustrations. This dictionary is suitable for the undergraduate population and the more sophisticated layperson. See entry 295 for the paperback edition. [R: Nature, Dec. 1984, p. 675]

279. **Flowering Plants of the World**. Edited by V. H. Heywood, et al. Englewood Cliffs, N.J.: Prentice-Hall, 1985. 335p. ill. maps. index. $39.95. ISBN 0133224058.

Flowering Plants of the World is written and prepared by a prestigious group of international authorities. The aim of this reference work is to appeal to both the general public and the professional, and it succeeds admirably in this goal. The comprehensive survey of the Angiospermae provides information on flowering plants in general, their classification, structure, contributions, and terminology. The book is arranged systematically to include 250,000 species in 300 flowering plant groups. Each family receives expert attention regarding its characteristics, distribution, diagnostic features, classification, and economic uses. [R: ARBA, 1986, entry 1497]

280. Gerth Van Wijk, H. L. **Dictionary of Plant Names**. rev. ed. The Hague: Nijhoff, 1911-16; repr., Monticello, N.Y.: Lubrecht and Cramer, 1971. 2v.

Volume 1 includes a bibliography and an alphabetical list of scientific names with extensive lists of corresponding English, French, German, and Dutch vernacular names. Volume 2 is the index to vernacular names. This monumental work is unique and accurate.

281.   Graf, Alfred Byrd. **Exotica. Series 4 International: Pictorial Cyclopedia of Exotic Plants from Tropical and Near-tropic Regions**. 11th ed. East Rutherford, N.J.: Roehrs, 1985. 2v. $187.00/set. ISBN 0911266208.

This contemporary pictorial record of ornamental or fruited plants and trees is an exhaustive encyclopedia, presenting 16,300 photographs, 405 plants in color, and 300 drawings. This is a guide to care, decorating, gardening in the tropics, plant geography and ecology, and propagation. There are discussions for each plant on its origins, botanical synonyms, cross-references, common name, and directions for care in English, Spanish, German, French, and Russian. For authors responsible for naming the plant, consult *Hortus III* (entry 283).

282.   Grebenshchikov, Oleg S. **Geobotanic Dictionary: Russian-English-German-French**. Moscow: Nauka, 1965; repr., Forestburgh, N.Y.: Lubrecht and Cramer, 1979. 226p. $31.50. ISBN 3874291642.

Geobotanical terminology, principal plant formations and world plant communities, and related terms from plant ecology, soil science, climatology, geomorphology, and phytogeography are included in the 2,660 entries. The contents include a Russian-English-German-French dictionary with English, German, and French indexes. This useful dictionary can be supplemented and updated by Elsevier (entry 272) and Davydov (entry 268).

283.   **Hortus III: A Concise Dictionary of Plants Cultivated in the United States and Canada**. Compiled by Liberty H. Bailey and Ethel Z. Bailey. Revised and expanded by the staff of the Liberty Hyde Bailey Hortorium. New York: Macmillan, 1976. 1290p. ill. $125.00. ISBN 0025054708.

As stated in the preface, this is an "inventory of accurately described and named plants of ornamental economic importance." This is an indispensable source for description, botanical name and synonyms, common names, uses, propagation, hardiness, and illustrations. A glossary and index to common names are included. Updated by the *New York Botanical Garden Illustrated Encyclopedia of Horticulture* (see entry 293).

284.   Howes, Frank N. **A Dictionary of Useful and Everyday Plants and Their Common Names**. New York: Cambridge University Press, 1974. 300p. $39.50. ISBN 0521085209.

This dictionary, based on material contained in Willis's *A Dictionary of the Flowering Plants and Ferns* (6th ed., 1931), includes general botanical information concerned with economic uses, ornamental plants, and common names that were left out of Willis's seventh and eighth editions. This dictionary definitely serves a useful purpose by itself and in conjunction with later editions of Willis (see entry 310). [R: ARBA, 1975, entry 1463; LJ, Dec. 1, 1974, p. 3124]

285.   Jackson, Benjamin D. **A Glossary of Botanic Terms with Their Derivation and Accent**. 4th ed. rev. and enl. New York: Hafner, 1960. 481p. $27.50. ISBN 0028471105.

This well-known glossary has 25,000 entries providing derivation of terms, accent, and definition. It puts terms into context with the use of their period, and gives authors' names (in parentheses) for authority of the definition. This dictionary may well be used in conjunction with one that is more modern; however, the information that Jackson provides is unique in many ways and is especially useful for historical work.

286.   Jaeger, Edmund C. **A Source-Book of Biological Names and Terms**. 3rd ed. Springfield, Ill.: Charles C. Thomas, 1978. 360p. ill. $35.50. ISBN 0398009163.

This is larger, more current, and much more useful than Borror (see entry 262). There are concise discussions on the building of words, types of names considered, transliteration, Greek prefixes, the form of Latin nouns, and over 280 brief biographies of people commemorated in botanical and zoological generic names.

287.   Kadans, Joseph M. **Modern Encyclopedia of Herbs with the Herb-o-matic Locator Index**. Englewood Cliffs, N.J.: Prentice-Hall, 1970. 256p. $4.95pa. ISBN 0135937809.

Herbs are arranged by their common names; botanical names, effects, and reported usage are included. There is a section on spice and herb cookery. The herb-o-matic index lists herbs, symptoms, health situations, and uses. This and Morton (entry 292) are complementary.

288.   Little, R. John, and C. Eugene Jones. **Dictionary of Botany**. New York: Van Nostrand Reinhold, 1983. 400p. ill. $14.95pa. ISBN 0442260199.

Approximately 5,500 concise definitions from all fields of botany are included in this useful dictionary; it is appropriate for all levels of expertise. Common and taxonomic names are not included, nor are structural formulas for chemical compounds. Every effort was made to locate and include new botanical terms from varied sources. Black-and-white drawings illustrate the definitions and a bibliography is included.

289.   **Longman Illustrated Dictionary of Botany**. By Andrew Sugden. New York: Longman, 1984. 196p. ill. (col.). $7.95. ISBN 0582556961.

"The elements of plant science illustrated and defined." This well-illustrated and colorful little dictionary is useful for definitions and synonyms. It is arranged by subject area and contains over 1,200 words used in the botanical sciences. Its main features are the many color illustrations averaging more than one per page. It is current and a good choice for the beginner or layperson. [R: Nature, Dec. 13, 1984, p. 675]

290.   Miller, Paul R., and Hazel L. Pollard. **Multilingual Compendium of Plant Diseases**. St. Paul, Minn.: American Phytopathological Society for the United States Agency for International Development in cooperation with the U.S. Department of Agriculture, 1976-77. 446p. $40.00. ISBN 0890540209. (Compendia Series, Vol. 2).

Although plant pathology is generally out-of-scope for this guide, this particular compendium is included as a communication aid for botanists working with translations. The entries are arranged by Latin name of the host in association with the Latin name of the pathogen, resulting in the name of the disease in English and 20 other languages. Each disease is described in English, Interlingua, French, and Spanish; 22 language indexes refer to the disease descriptions.

291.   Moerman, Daniel E. **American Medical Ethnobotany: A Reference Dictionary**. New York: Garland, 1977. 527p. $66.00. ISBN 0824099079. (Garland Reference Library of Social Science Series, Vol. 34).

This book serves as a guide to the native American medicinal uses of plants and the literature that supports its study. There are tables for genera, indications, families, and cultures. Each table lists all the items in the databank: 1,288 plant species from 531 different genera used in 48 different cultures in 4,869 different ways. All entries refer to one of 41 sources; there is a common name index and a supplementary bibliography. This is a valuable source for anyone interested in the medicinal use of plants by American Indians.

292.   Morton, Julia Frances. **Herbs and Spices**. New York: Golden, 1977. 160p. ill. (col.). $2.95pa. ISBN 0307243648. (Golden Nature Guides).
       This reference work, written by a recognized authority, discusses 372 worldwide species of wild and cultivated flavoring plants. There is a general introduction to herbs and spices, information on herb gardening, harvesting, preserving, and culinary uses. An excellent color illustration is provided for each plant. The book is arranged by family and includes information on common and botanical names, distribution, cultivation, description, and uses. There is an index.

293.   **New York Botanical Garden Illustrated Encyclopedia of Horticulture**. Compiled by Thomas H. Everett. New York: Garland, 1980-82. 10v. $600.00/set. ISBN 0824072227.
       This beautiful and outstanding reference source supplements *Hortus III* (entry 283). Although it is intended for quick reference on horticulture for gardeners, it is extremely useful for botanists of all kinds. It is comprehensive and authoritative, containing major botanical articles. The encyclopedia includes 3,601 pages, 10,000 photographs, and descriptions of 20,000 species and varieties of plants.

294.   **Oxford Encyclopedia of Trees of the World**. Bayard Hora, ed. consultant. New York: Oxford University Press, 1981. 384p. ill. (col.). $27.50. ISBN 0192177125.
       Oxford Press encyclopedias have a good reputation and this one does not disappoint. This attractive and authoritative book covers a broad field in depth to provide a summary of the major genera of trees of the world. Beautiful color photographs and illustrations cover trees of every kind, with information given on the identification of trees. This book is an excellent general source for native and exotic trees of North America. A bibliography, glossary, and common and Latin name indexes are included. [R: RQ, Summer 1982, p. 420]

295.   **Penguin Dictionary of Botany**. By Lawrence Urdang and Associates. Edited by Stephen Blackmore and Elizabeth Tootill. New York: Penguin, 1984. 288p. $7.95pa. ISBN 0140511261. See entry 278.

296.   Perry, Frances. **Flowers of the World**. Illustrated by Leslie Greenwood. New York: Crown, 1972. 320p. ill. (col.). $22.50. ISBN 060001634X.
       This beautiful book covers selected examples of the major families of tropical, subtropical, and temperate plants. Its purpose is to outline the "vast wealth of plant species throughout the world" and it does this very handsomely, with 828 illustrations in full color. There are a small glossary and a bibliography; the bulk of the book is given over to readable descriptions and the abundant colored illustrations of the flowers of the world. An index for common and scientific names is included. [R: Choice, Mar. 1973, p. 64]

297.   Plowden, Chicheley C. **A Manual of Plant Names**. 2nd ed. London: Allen and Unwin, 1970. 260p.
       This dictionary is appropriate for the beginning student or hobbyist. Its purpose is to bring awareness and enjoyment of plant names and their meanings to the layperson.

There are sections discussing generic names, specific epithets, common names, botanical terms, and the leaf and plant system.

298. **Popular Encyclopedia of Plants**. Edited by Vernon H. Heywood and Stuart R. Chant. New York: Cambridge University Press, 1982. 368p. ill. (col.). $34.50. ISBN 0521246113.

In addition to the alphabetized section, there are 21 special feature articles on economic plant groups or products. Over 700 color photographs illustrate 800 species of economically important plants and several hundred ornamental species. One or more paragraphs are devoted to each entry, giving information on form and structure, reproduction, distribution, ecology, and uses. This attractive, very readable encyclopedia is appropriate for professional botanists or the general reader. There are indexes for common and scientific names and a bibliography is included. [R: CRL, Jan. 1983, p. 53]

299. Shimokoriyama, Masami. **Saishin shokubutsu yogo jiten/A Dictionary of Botanical Terms**. 9th ed. Tokyo: Hirokawa, 1982. 679p.

Japanese dictionary of botanical terms provides terms in Japanese, English, German, French, and Latin.

300. Shosteck, Robert. **Flowers and Plants: An International Lexicon with Biographical Notes**. New York: Quadrangle/New York Times Book Company, 1974. 329p. ISBN 0812904532.

This dictionary is arranged by common name of flora in North America and Canada. There are several thousand entries that include botanical name, origin, use, and history for each plant. The accounts are interesting, enjoyable reading, with at least one drawing per page. A glossary, bibliography, and index are provided. Biographical sketches of notable figures in botanical history are included, making this dictionary, along with Jaeger (entry 286), important to the historian. [R: ARBA, 1975, entry 1464]

301. Smith, Archibald William. **Flowering Plants of the World**. Vernon H. Heywood, ed. consultant. New York: Smith Publishers, 1978. 335p. $17.95. ISBN 0831734000.

This encyclopedic treatment presents an introduction to forms, structure, ecology, uses, and classification of the flowering plants of the world and includes a glossary. More than 300 Angiosperm families are described and illustrated, many in full color. The descriptions include information on distribution, significant features, classification, economic uses, number of species and genera. The work was written by internationally recognized authorities on the subject and is scientifically accurate.

302. Smith, Archibald William. **Gardener's Dictionary of Plant Names: A Handbook on the Origin and Meaning of Some Plant Names**. Revised and enlarged by W. T. Stearn. New York: St. Martin's Press, 1971. 391p.

The purpose of this book is to provide a source of reliable information for gardeners on the significance of botanical names attached to plants. W. T. Stearn includes a fascinating essay on botanical names, with a short bibliography for sources of additional reading. Vernacular names are introduced in a similar fashion, keeping the dictionary informative as well as interesting.

303. Snell, Walter H., and Esther A. Dick. **A Glossary of Mycology**. rev. ed. Cambridge, Mass.: Harvard University Press, 1971. 181p. $15.00. ISBN 0674354516.

Mycological terms and related entries useful to the student of mycology make up this dictionary. For the most part, the definitions are brief and include accurate, authoritative etymological information. This is a reasonable alternative to *Ainsworth and Bisby's* (entry 258) for definition of terms, but it is not nearly as comprehensive. It does include color terms, relating them to appropriate equivalents from two-color standards of importance to mycologists.

304. Stearn, William T. **Botanical Latin: History, Grammar, Syntax, Terminology, and Vocabulary**. 2nd ed. North Pomfret, Vt.: David and Charles, 1983. 566p. ill. $32.00. ISBN 0715385488.
"A guide to the special kind of Latin internationally used by botanists for the description and naming of plants," this volume is included in the dictionary section because of its usefulness with other, more conventional dictionaries. It traces the development of botanical Latin terminology, introducing the Latin alphabet, pronunciation, grammar, and usage to the uninitiated. There is information on writing botanical diagnoses and a third of the guide is devoted to a Latin-English, English-Latin vocabulary for terms and expressions relating to plants. This is an authoritative, scholarly treatment.

305. Steinmetz, E. F. **Codex Vegetabilis**. 2nd ed. Amsterdam: Published by the author, 1957. 136p.
Medicinal plants in English, Dutch, German, French, Italian, Danish, Spanish, Latin, and other European and Oriental languages. Arranged by botanical name, 1,216 entries provide information on habitat, useful parts, constituents, action, common names in English, Dutch, and German, etc. There is an index including all names and languages.

306. Steinmetz, E. F. **Vocabularium botanicum. Planten terminologie. Woordenlijst in zes talen (Latin, Grieks, Nederlands, Duits, E Engels, Frans) van de voornaamste wetenschappelijke woorden, die in de plantkunde gebruikt worden**. 2nd ed. Amsterdam: Published by the author, 1953. 149p.
This is another equivalents dictionary useful for botanical terms.

307. Swartz, Delbert. **Collegiate Dictionary of Botany**. New York: Ronald, 1971. 520p.
Nearly 24,000 entries are included in this botanical dictionary compiled from more than 170 sources. General botanical terms are not included, with American usage rather than British. An outline of the plant kingdom is provided at the end of the book. Where necessary for comprehensiveness, several definitions are provided for a single term. In general the definitions are short, uninvolved, and clear. [R: Phyto, Feb. 1973, p. 264]

308. **Tanaka's Cyclopedia of Edible Plants of the World**. By Tyozaburo Tanaka. Tokyo: Keigaku, 1976. 924p.
This dictionary includes more than 10,000 edible plants with Latin, family, common names, uses, edible portions, distribution, synonyms, and references to earlier work in the literature. There is an extensive bibliography and a comprehensive index.

309. Uphof, Johannes Cornelius Theodor. **Dictionary of Economic Plants**. 2nd ed. Monticello, N.Y.: Lubrecht and Cramer, 1968. 591p. $28.00. ISBN 3768200019.

This technical dictionary provides short descriptions of 9,500 species of economic plants. Geographical distribution, products, and principal uses are provided; there is a comprehensive bibliography arranged by subject.

310.   Willis, John Christopher. **A Dictionary of the Flowering Plants and Ferns**. 8th ed., rev. New York: Cambridge University Press, 1973. 1245p. 1985 pa. ed. $39.50. ISBN 0521313953.

This is the standard taxonomic dictionary of the flowering plants and ferns. It is authoritative, comprehensive, and includes a key to the families of flowering plants based on Engler's *Syllabus* (ed. 12), the system of Bentham and Hooker, and a table of family equivalents for this edition of the dictionary. Every published generic name from 1753 on, and every published family name from the appearance of the *Genera Plantarum* of Jussieu in 1789, is included. All of the entries deal with taxonomic relationships; entries for families are descriptive and detailed.

5 Handbooks and Methods and Techniques Books

This chapter is divided into two parts, handbooks, and methods and techniques. Handbooks, which will be considered first, include those reference materials that provide data or information in a compact form conveniently arranged for easy access. The information is usually up-to-date, comprehensive, and based on authoritative sources. Handbooks discussing procedures are included in the following section. Identification handbooks/manuals are included in chapter 7, "Identification Sources." The scope of the first section is wide ranging to cover aspects of microscopy, genetic maps, color manuals, elements of writing style, chemistry, physics, and other topics as they are relevant to botany.

Handbooks

311.   Bracegirdle, Ian, and Philip H. Miles. **An Atlas of Plant Structure**. Portsmouth, N.H.: Heinemann Educational Books, 1971. 2v. Vol. 1: $20.00. ISBN 0435603124. Vol. 2: $20.00. ISBN 0435603140.

This set was developed to help students interpret their laboratory specimens, and to this end, each photomicrograph example is faced by an interpretive line drawing.

Bacteria, algae, fungi, lichenes, hepaticae, musci, and other plant tissues are included for examination and comparison. This source is unique and of great assistance in the laboratory.

312. **Catalogue of the Culture Collection of the Commonwealth Mycological Institute.** 8th ed., including 1983 supplement. Kew, England: Commonwealth Agricultural Bureaux, 1982. 224p. $31.50. ISBN 0851985025.

This valuable catalog lists the CMI Culture Collection, which incorporates the United Kingdom National Collection of Fungus Cultures, a public collection and an integral part of the Institute's identification service. The collection includes more than 10,000 isolates of interest in plant pathology, industry, biodeterioration studies, taxonomy, and biochemical research and education, with the greatest emphasis placed on Phycomycetes, Ascomycetes, and Fungi Imperfecti. The catalog provides information on maintenance, ordering, exchange and deposit, and other services of CMI. The culture list occupies the bulk of the catalog with appendixes on cultures for testing, teaching, bacteria, and a metabolism index. Information provided for cultures includes names, synonyms, accession number and additional deposits, number of isolates from the same host, type material, host material, country of origin and date, name of mycologist who isolated the fungus, dates of workers, and annotations.

313. **CBE Style Manual: A Guide for Authors, Editors, and Publishers in the Biological Sciences.** 5th ed., rev. and exp. Prepared by the CBE Style Manual Committee. Bethesda, Md.: Council of Biology Editors, 1983. 324p. $26.75. ISBN 0914340042.

The standard style manual for the biological sciences provides a plant sciences section in the chapter, "Style in Special Fields." This manual discusses ethics, manuscript preparation, editorial review, copyright, publishing, style conventions, and secondary sources; there is an annotated bibliography and an index.

314. **CRC Handbook of Biochemistry and Molecular Biology.** 3rd ed. Edited by G. D. Fasman. Boca Raton, Fla.: CRC Press, 1976. 8v. price varies. (Vol. 1: $76.50. ISBN 0878195084). **Cumulative Series Index.** $56.00. ISBN 084930511X.

This set is divided into sections for proteins; nucleic acids; lipids, carbohydrates, and steroids; and physical and chemical data.

315. **CRC Handbook of Chemistry and Physics.** 67th ed. Edited by Robert C. Weast. Boca Raton, Fla.: CRC Press, 1986. 2424p. $69.95. ISBN 0849304670.

Revised annually. A classic "must" for phytochemists providing reliable and authoritative chemical and physical data.

316. **CRC Handbook of Flowering.** Edited by Abraham H. Halevy. Boca Raton, Fla.: CRC Press, 1985- . 5v. $800.00/set. Vol. 1: ISBN 0849339111. Vol. 2: ISBN 084933912X. Vol. 3: ISBN 0849339138. Vol. 4: ISBN 0849339146.

This reference aims to be comprehensive for information on the control and regulation of flowering. The projected set will provide more than 2,000 pages of information encompassing over 5,000 species of plants with specific data on all aspects of flower development, sex expression, requirements for flower initiation and development, photoperiod, light density, vernalization, and other temperature effects and interactions. The set should be useful in both applied and theoretical areas for a variety of students, researchers, scientists, laboratories, and libraries.

317. Culberson, Chicita F. **Chemical and Botanical Guide to Lichen Products**. Chapel Hill, N.C.: University of North Carolina Press, 1969. 628p. **Supplements, 1-2**. Springfield, Mo.: American Bryological and Lichenological Society, 1970-77. $18.00/set.

The purpose of this guide and its supplements is to summarize the literature on the nature and occurrence of lichen substances.

318. **Genetic Maps 1984: A Compilation of Linkage and Restriction Maps of Genetically Studied Organisms**. Edited by Stephen J. O'Brien. Cold Spring Harbor, N.Y.: Cold Spring Harbor Laboratory, 1984. 583p. $28.00. ISBN 0879691719.

Completely revised and updated, this edition contains 85 maps and covers 75 organisms. Contents of interest to botanists include data for algae, fungi, and selected plants. There are also complete lists of materials contained in the following nucleic acid and protein sequence databases: National Biomedical Research Foundation, Genbank (NIH), and the European Molecular Biology Laboratory (EMBL).

319. **Handbook of Natural Toxins**. Edited by Richard J. Keeler and Anthony T. Tu. New York: Marcel Dekker, 1983- . 5v. (Vol. 1: $145.00. ISBN 0824718933).

This set includes the following: volume 1: Plant and fungal toxins; volume 2: Insect poisons, allergens, and other invertebrate venoms; volume 3: Bacterial toxins; volume 4: Reptile and amphibian venoms; and volume 5: Marine venoms and toxins. Volume 1 includes 24 papers from 38 expert contributors that collate and interpret results from field observations and laboratory investigations on plant and fungal poisoning. Toxins are grouped by effects on cardiovascular or pulmonary systems, carcinogenic effects, reproductive effects, psychic or neurotoxic effects, gastrointestinal or hepatic effects, effects on species interactions, and usefulness in medicine. The text provides contemporary information on chemistry, source, gross and histopathologic effects, and the mechanism of action. There are author and subject indexes. This is a useful resource for both research and applied areas.

320. Kelly, Kenneth Low, and Deane B. Judd. **Color: Universal Language and Dictionary of Names**. Washington, D.C.: U.S. Department of Commerce, National Bureau of Standards; distr., Washington, D.C.: Government Printing Office, 1976. 158p. $3.25. (NBS Special Publication, 440).

This publication supersedes and combines *The ISCC-NBS Method of Designating Colors and a Dictionary of Color Names*, by Kelly and Judd, and *A Universal Color Language* by Kelly. It provides a standardized color nomenclature with excellent color charts to provide accuracy and convenience in one method for color determination.

321. **Microbiological Resource Databank (MIRDAB) Catalog 1/1985**. New York: Elsevier Scientific Publishing, 1985. 612p. $83.50. ISBN 0444903879.

This printed version of the MIRDAB databank contains 601 records of animal cells, plant cells, and animal viruses. The information gives specific data on the cell or virus, its source and availability, an entry to the published scientific literature, and indexes to cell/virus names and applications. Eventually, this catalog will become a computerized databank when the mass of collected data warrants it.

322. Mitchell, John, and Arthur Rook. **Botanical Dermatology: Plants and Plant Products Injurious to the Skin**. Vancouver, Canada: Greenglass, 1979. 787p. ISBN 0889780471.

This encyclopedic and authoritative reference includes a short history of injurious plants with references to the literature. Irritant and allergic plants from 1405 genera in 248 families are arranged alphabetically, including brief discussions and citations to the literature for their chemistry, dermatological reactions, veterinary aspects, and patch testing. Refer to Lampe (entry 511) for additional or supplementary information.

323. **Mycology Guidebook**. Edited by Russell B. Stevens. Published under the auspices of the Mycology Guidebook Committee, Mycological Society of America. Seattle, Wash.: University of Washington Press, 1981. 736p. $40.00. ISBN 0295958413.

This indispensable manual assembles information to assist teachers in preparing mycological laboratory courses, but its usefulness is much broader. It provides information on field collecting, isolation techniques, culture maintenance, taxonomic groups, ecological groups, fungi as biological tools, and extensive references for culture repositories, stains and media, and available films.

324. **Plant Information Network (PIN) Database: Colorado, Montana, North Dakota, Utah, and Wyoming**. Compiled by Phillip L. Dittberner and Michael R. Olson. Performed for Western Energy and Land Use Team, Division of Biological Services, Research and Development, Fish and Wildlife Service, U.S. Department of the Interior. Washington, D.C., 1984. 786p. microfiche. FWS/OBS-83/86.

Although this handbook was of regional origin, its uses are far broader to store, organize, and retrieve information on 5,000 native and naturalized plants of several western states. Basically arranged in tabular form under headings of taxonomy, geography, biology, ecology, and economic plant attributes, this hard copy of the PIN database (which is no longer available because of funding problems) gives pertinent data relating to each plant listed. There is an extensive bibliography. Applications include environmental impact statements, vegetation inventory, reclamation planning, ecological research, distribution, and habitat data on endangered plants.

325. Prance, Ghillean Tolmie. **Leaves: The Formation, Characteristics and Uses of Hundreds of Leaves Found in All Parts of the World**. New York: Crown Publishers, 1985. 244p. ill. (col.). index. $35.00. ISBN 0517551527.

This book has been described as "the best presentation of the subject." It is well written, brief, and directed toward the lay reader without being too simplistic for the informed. It covers leaf anatomy, physiology, and design; useful leaves; patterns of leaf damage; carnivorous, poisonous, and fossil leaves; and collecting and leaf printing. The color photographs by Kjell B. Sandved are especially beautiful. It is "highly recommended." [R: ARBA, 1986, entry 1488; LJ, May 1, 1985, p. 69]

326. Raffauf, Robert Francis. **A Handbook of Alkaloids and Alkaloid-containing Plants**. New York: Wiley Interscience, 1970. 1v.

This reference is a compilation of data concerning plant alkaloids and their distribution. The body of the work is a set of tables giving the plant source, formulas, and physical properties of the known alkaloids arranged in alphabetical order of the plant families in which they occur. Data include name and structure, botanical family and genus of origin, molecular formula, molecular weight, melting point, optical rotation, and literature citation.

327. Ridgway, Robert. **Color Standards and Color Nomenclature**. Washington, D.C.: Published by the author, 1912.

This classic color system includes 53 colored plates naming 1,115 colors. It has been the color standard for biologists since its publication and is still heavily referred to by botanists although it has been out of print for a long time.

328.   Smithe, Frank B. **Naturalist's Color Guide**. Pts. 1-3. New York: American Museum of Natural History, 1974-81. (Pt. 3: $8.00. ISBN 0913424056).

A contemporary and very well-done color system, designed by Smithe to rigorous standards, includes 86 colors with names and numerical notation of color measurement. The colors have been carefully controlled and can be specifically identified, using Ridgway's terminology whenever possible.

Methods and Techniques

This selection of works on methods and techniques concentrates on recent material and continuing series of importance to the student and researcher. The scope of this section encompasses very technically oriented chemical and physical methods, as well as techniques of interest to the amateur botanical artist. Laboratory manuals, as such, are not included because emphasis is placed on understanding principles and concepts rather than presenting a cookbook of procedures. Clinical reference works are included only as examples or references, leading to more comprehensive works from the medical field. The most current information on methods, of course, can be found in the primary journal literature and in proceedings of recent conferences and symposia; for access to these materials, see chapter 2.

329.   Abbott, Lois A., Frank A. Bisby, and David J. Rogers. **Taxonomic Analysis in Biology: Computers, Models, and Databases**. New York: Columbia University Press, 1985. 336p. ill. bibliog. index. $40.00. ISBN 0231049269.

Written for those with no experience in taxonomy, this useful guide describes process and methods of taxonomy across the field of biology including (1) standard methods used traditionally, (2) formal aspects of data structure and math models, (3) practical details for computer-assisted taxonomic analysis, and (4) practical means of data storage and information retrieval with examples.

330.   **Analytical Methods for Pesticides, Plant Growth Regulators, and Food Additives**. Vol. 1- . Edited by Gunter Zweig and Joseph Sherma. New York: Academic, 1968- . (Vol. 13: 1984. $71.50. ISBN 0127843132).

This continuing series, appropriate for the botanical library/laboratory, is useful for students and researchers in applied areas of the plant sciences.

331.   Berlyn, Graeme P., and Jerome P. Miksche. **Botanical Microtechnique and Cytochemistry**. Ames, Iowa: Iowa State University Press, 1976. 326p. ill. bibliog. index. $16.50. ISBN 0813802202.

This manual, a revision of the well-known *Botanical Microtechnique* by J. E. Sass, can be used to introduce teachers and researchers in plant science to basic principles of microtechnique and cytochemistry. It is not an inventory of techniques, but it provides a good introduction to some of the methods of interest to botanists. The bibliography is helpful. [R: Choice, Oct. 1976, p. 1004]

332. **Botanical Microscopy, 1985**. Edited by Anthony W. Robards. New York: Oxford University Press, 1985. 368p. bibliog. index. $29.95. ISBN 0198545878.

This book presents the invited papers from the Third International Meeting on Botanical Microscopy, held in York in July, 1985, under the auspices of the Royal Microscopical Society. Papers from leading experts discuss state-of-the-art research techniques of interest to botanists, including scanning and transmission electron microscopes, light microscopy, image enhancement techniques, fluorescence, and microinjection methods.

333. Campbell, Mary C., and Joyce L. Stewart. **The Medical Mycology Handbook**. New York: John Wiley, 1980. 454p. ill. (col.). $36.00. ISBN 0471047287.

Although the emphasis is medical, this is extremely useful for any mycological laboratory worker. Taxonomy, fungal disease, clinical laboratory methods, information on lab equipment, reagents, media, stock culture maintenance, and informative color plates make this practical bench handbook of assistance for students and researchers.

334. **Cell Culture and Somatic Cell Genetics of Plants**. Vol. 1- . New York: Academic, 1984- . (Vol. 1: 1984. $93.50. ISBN 0127150013).

This treatise on plant cell culture is designed to provide key reference works, descriptions, and discussions of all aspects of modern plant cell and tissue culture techniques and research. It is written by specialists and authorities in the field, and is useful for either the novice or the experienced researcher. Volume 1, edited by Indra K. Vasil, *Laboratory Procedures and Their Applications*, presents helpful how-to instructions with a good selection of important references in an attractive layout and text. This set is a competitor in many ways to the *Handbook of Plant Cell Culture* (entry 348), but the differences in subject matter, style, and content warrant its purchase. [R: Nature, Jan. 17, 1985, pp. 251-52]

335. Collins, Christopher Herbert, and Patricia M. Lyne. **Microbiological Methods**. 5th ed. Stoneham, Mass.: Butterworths, 1984. 448p. $34.95. ISBN 040870957X.

Although this is primarily useful for bacteriology, there are three chapters on yeasts, molds, and pathogenic fungi. The discussions on laboratory safety and equipment, sterilization, culture media, staining, and mycological methods are relevant and important for the student.

336. Darlington, Cyril Dean, and L. F. La Cour. **The Handling of Chromosomes**. 6th ed. Revised by L. F. La Cour. New York: John Wiley, 1975. 208p. ill. ISBN 0045740143.

This is an authoritative selection of methods for handling chromosomes for teaching and research. See Sharma and Sharma (entry 366) for additional and more recent information.

337. Gahan, Peter B. **Plant Histochemistry and Cytochemistry**. New York: Academic, 1984. 301p. ill. bibliog. index. $46.00. ISBN 0122732707. (Experimental Botany, Vol. 18).

This work presents basic methods for cytochemical analysis of plant cells and tissues.

338.   Geary, Ida. **Plant Prints and Collages**. New York: Viking Press, 1975. 101p. ISBN 0670558907.

This little book on nature printing will be interesting to public library readers and amateur botanical artists wishing to learn how to make prints with plants. There is a brief history of plant prints, information on methods for printing on cloth and decorative papers, information on how to press and dry plants and algae, and a short reading list and directory of sources for material. The book is not technical and presents a "pleasant introduction to an interesting set of techniques." [R: QRB, Sept. 1979, p. 336]

339.   **Genetic Engineering: Principles and Methods**. Vol. 1- . Edited by Jane K. Setlow and Alexander Hollaender. New York: Plenum, 1979- . irregular. $50.00. ISSN 0196-3716.

This series deals with the new technology of genetic engineering, its techniques, principles, and methods. Unlike the series *Genetic Engineering* published by Academic Press, this series includes information of use to those interested in plant techniques.

340.   George, Edwin F., and Paul D. Sherrington. **Plant Propagation by Tissue Culture: Handbook and Directory of Commercial Laboratories**. Hampshire, England: Exegetics, Ltd, 1984. 709p. bibliog. indexes. £65.00. ISBN 0950932507.

This book is included here because of its useful information on methods for plant tissue culture research and commercial laboratories. [R: QRB, Sept. 1985, p. 360]

341.   Glimn-Lacy, Janice, and Peter B. Kaufman. **Botany Illustrated: Introduction to Plants, Major Groups, Flowering Plant Families**. New York: Van Nostrand Reinhold, 1984. 146p. ill. bibliog. index. $19.45. ISBN 0442229690.

This book is written for students of botany and botanical illustration who are interested in learning more about plants with reference to methods of illustration. The book is divided into three parts: (1) an introduction to botanical facts; (2) a section on major groups from bacteria to flowering plants; and (3) a section on major flowering plant families. Each page is a separate subject with a separate full page of illustrative examples of scientifically accurate line drawings including a coloring code guide. There are brief explanations about methods and suggestions for sources of plants. This is an unusual book that is included here because of its discussion about drawing materials and for its excellent examples of illustration technique.

342.   **A Growth Chamber Manual: Environmental Control for Plants**. Edited by Robert W. Langhans. Ithaca, N.Y.: Comstock Publishing Associates, 1978. 222p. ISBN 0801411696.

This useful book serves as an introduction for the new user and as a reference for technicians, researchers, and students. The book explains methodology, management, and operation of growth chambers, and includes a list of manufacturers. [R: QRB, June 1979, p. 192]

343.   Haley, Leanor D., and Carey S. Callaway. **Laboratory Methods in Medical Mycology**. 4th ed. Atlanta, Ga.: U.S. Department of Health, Education and Welfare, Public Health Service Bureau of Laboratories, Laboratory Training and Consultation Division, 1978. 225p. (HEW Publication No. CDC 78-8361).

This successful methods book presents procedures for isolation and identification of fungi. Both Barnett (entry 460) and McGinnis (entry 518) are additional sources for consultation in this field of study.

344. **Handbook of Phycological Methods: Culture Methods and Growth Measurements.** Edited by Janet R. Stein. New York: Cambridge University Press, 1973. 512p. ill. $59.50; $21.95pa. ISBN 0521200490; 0521297478pa.

Sponsored by the Phycological Society of America, this compendium of techniques may be "profitably utilized by investigators at all levels." [R: Choice, Mar. 1974, p. 120]

345. **Handbook of Phycological Methods: Developmental and Cytological Methods.** Edited by Elisabeth Gantt. New York: Cambridge University Press, 1980. 425p. $44.50. ISBN 0521224667.

This second handbook sponsored by the Phycological Society of America presents well-organized sections on the isolation of organelles and membranes, analysis of chemical membranes, enzymes, physiological and biochemical processes, nutrients, ion content and transport, inhibitors, and suppliers.

346. **Handbook of Phycological Methods: Ecological Field Methods, Macroalgae.** Edited by Mark M. Littler and Diane S. Littler. New York: Cambridge University Press, 1986. 617p. $49.50. ISBN 0521249155.

This book, volume 4 of this handbook series, is the first comprehensive treatment of recently developed methodologies in marine benthic algal ecology. Both traditional and modern methods are included, along with information on adapting procedures for different habitats, diverse algal systems, or other conditions.

347. **Handbook of Phycological Methods: Physiological and Biochemical Methods.** Edited by Johan A. Hellebust and J. S. Craigie. New York: Cambridge University Press, 1978. 512p. $57.50. ISBN 0521218551.

The third volume in this four-volume set sponsored by the Phycological Society of America presents methods for physiological and biochemical investigations with algae. The book is well written, "a must" for both research workers and instructors. [R: QRB, Sept. 1981, p. 342]

348. **Handbook of Plant Cell Culture.** Vol. 1- . New York: Macmillan, 1983- . (Vol. 1: 1983. $53.00. ISBN 0029492300).

This projected five-volume set aims to present a unified summary of plant cell culture techniques. It provides a comprehensive, critical review of contributions from the literature to focus on techniques, detailed protocols, and applications of cell culture. The reviewer in *Nature* calls it "generally successful" in its attempt to produce a definitive statement, an "invaluable reference work." This set, and its competitor *Cell Culture and Somatic Cell Genetics of Plants* (entry 334), are both highly recommended. [R: Nature, Jan. 17, 1985, pp. 251-52]

349. Hangay, George, and Michael Dingley. **Biological Museum Methods.** Orlando, Fla.: Academic, 1985. 2v. Vol. 1: **Vertebrates.** 400p. $75.00. ISBN 0123233011. Vol. 2: **Plants, Invertebrates, and Techniques.** 352p. $58.00. ISBN 012323302X.

This two-volume set describes methods of collecting, preparing, and displaying biological material for museums, universities, colleges, and schools. Plants and relevant techniques are included in volume 2.

350. Harborne, Jeffrey B. **Phytochemical Methods: A Guide to Modern Techniques of Plant Analysis**. 2nd ed. London: Chapman and Hall; distr., New York: Methuen, 1984. 300p. $33.00. ISBN 0412255502.

This authoritative reference is intended for students as a simple guide to recommended phytochemical techniques. Chapters describe methods for identifying phenolic compounds, terpenoids, organic acids and related compounds, nitrogen compounds, sugars and their derivatives, and macromolecules. Appendixes provide a checklist of thin layer chromatography procedures for all classes of plant substances and a list of useful addresses for rare chemicals, chromatographic equipment, and spectrophotometers.

351. Hill, Stephen A. **Methods in Plant Virology**. Oxford, England: Blackwell; distr., Palo Alto, Calif.: Blackwell Scientific Publications, 1985. 180p. ill. index. $24.00pa. ISBN 0632009950. (Methods in Plant Pathology, Vol. 1).

Published on behalf of the British Society for Plant Pathology, this methodology handbook presents basic techniques, rationale, materials required, step-by-step procedures, interpretation of results, additional references, and detailed recipes where appropriate. Basic biophysical and chemical knowledge is assumed. Chapters include histological and other basic methods, virus characterization and storage, transmission tests, serological techniques, and electron microscopy. Useful for the student at the bench, this volume also helps provide elementary understanding of plant virology.

352. Jastrzebski, Zbigniew T. **Scientific Illustration: A Guide for the Beginning Artist**. Englewood Cliffs, N.J.: Prentice-Hall, 1985. 336p. ill. bibliog. index. $49.95; $22.95pa. ISBN 0137959494; 0137959311pa.

Jastrzebski introduces the beginner to technical aspects of drawing and painting, covering process, basic techniques and tools, precise steps leading to a finished product, suggestions for projects, exercises, and tips on speciality areas.

353. **Journal of Tissue Culture Methods**. Vol. 1- . Gaithersburg, Md.: Tissue Culture Association, 1975- . quarterly. $50.00/yr. ISSN 0271-8057.

Formerly *TCA Manual*, this journal contains papers describing techniques developed during studies with cell, tissue, or organ cultures from multicellular animals or plants. Each technique or method is contributed by a specialist in the field. Beginning with volume 8, no. 4, three issues will contain procedures for specific areas of research and the fourth will contain contributed papers.

354. Kershaw, Kenneth A., and John Henry H. Looney. **Quantitative and Dynamic Plant Ecology**. 3rd ed. Baltimore, Md.: Edward Arnold, 1985. 282p. ill. bibliog. index. $27.50. ISBN 0713129085.

This new edition is aimed at the development of plant ecology, and to this end, the authors have added chapters on ordination and classification, with comparative analysis of two data sets as examples included in the appendixes. There is a reappraisal of succession and the section on allelopathy has been updated. Additional material on pattern analysis has been included with a worked data set in an appendix. Other material presented in the eight appendixes include variance ratios, correlation of coefficients, random numbers, and distribution of t and $x2$.

355. Knudsen, Jens W. **Collecting and Preserving Plants and Animals**. New York: Harper & Row, 1972. 320p. $16.95pa. ISBN 0060437448.

This book on biological techniques, a reduced version of the author's 1966 *Biological Techniques*, is appropriate for students or amateur botanists interested in collecting and working with living or prepared specimens of plants and animals. Four chapters are devoted to plants, while the rest pertain to zoology. For more recent material, see Robertson (entry 365). [R: NatHist, Oct. 1967, p. 66]

356. **Methods in Chloroplast Molecular Biology.** Edited by M. Edelman, R. B. Hallick, and N.-H. Chua. Amsterdam: Elsevier Biomedical Press, 1983. 1152p. $183.00. ISBN 0444803688.

This comprehensive methodological treatise discusses screening for mutants; chloroplast and subchloroplast isolation; preparation of plastid DNAs, ribosomes, mRNAs, and tRNAs; in vitro labeling of plastid nucleic acids; mapping by restriction endonuclease technology; constructing recombinant plastid DNAs; DNA-RNA hybridization techniques; in vitro transcription and translation; synthesis and transport of proteins; isolation of stromal, thylakoid, and envelope proteins; preparation of antibodies and immunochemical techniques; and procedures for analyzing plastid proteins.

357. **Methods in Plant Ecology.** Edited by S. B. Chapman. New York: Halsted, 1977. 589p. ISBN 0470992018.

This review of ecological methods, written by experts, includes complete syntheses of historical studies of vegetation, analyzing soils, etc. of interest to the beginning ecologist and the established professional. It is especially useful for its numerous citations to the literature and its emphasis on predominantly British work. This, Kershaw and Looney (entry 354), Mueller-Dombois (entry 359), and Orloci (entry 360) make a useful combination for plant ecologists. [R: QRB, Sept. 1977, p. 307]

358. **Modern Methods of Plant Analysis, New Series.** Vol. 1- . New York: Springer-Verlag, 1985- . (Vol. 1: 1985. 399p. $69.50. ISBN 0387158227).

This series aims to bring together and critically evaluate methods for biochemical research on plant materials. Volume 1, *Cell Components*, deals with the analysis of the plant cell into its component parts. All major cell components and organelles are covered and methods for analysis/preparation/use are indicated. Volume 2 is titled *Nuclear Magnetic Resonance* and volume 3 is *Chromatography/Mass Spectroscopy*.

359. Mueller-Dombois, Dieter, and H. Ellenberg. **Aims and Methods of Vegetation Ecology.** New York: John Wiley, 1974. 547p. ISBN 0471622907.

The aim of this volume is to combine "classical and current concepts with detailed prescriptions for vegetation analysis and data processing by well-proven methods." This synthesis of European and American approaches to vegetation science may be favorably compared and used with *Methods in Plant Ecology* (entry 357) and Orloci (entry 360).

360. Orloci, Laszlo. **Multivariate Analysis in Vegetation Research**. 2nd ed. The Hague: Junk; distr., Hingham, Mass.: Kluwer Academic, 1978. 451p. bibliog. indexes. $53.00. ISBN 9061935679.

Orloci provides statistical and mathematical methods for the plant ecologist. There is an introduction to concepts and procedures and a source for worked examples and useful computer routines. "Serious students of vegetation will find this book indispensable." [R: QRB, Mar. 1976, p. 129]

361. **Plant Biosystematics**. Edited by W. F. Grant. Orlando, Fla.: Academic, 1984. 674p. ill. $54.50. ISBN 012295680X.

Containing the proceedings of a four-day symposium sponsored by the International Organization of Plant Biosystematists, held at McGill University, Montreal, in July, 1983, this volume discusses new biosystematic methods for detecting biotic units. The work is important to plant biosystematists and relevant to population biologists, biogeographers, cytogeneticists, and plant reproductive biologists.

362. **Plant Cell Culture: A Practical Approach**. Edited by R. A. Dixon. Arlington, Va.: IRL Press, 1985. 250p. ill. index. $25.00pa. ISBN 0947946225. (Practical Approach Series).

This handbook discusses methods for the establishment and manipulation of cell culture systems, plus brief general reviews of important culture variables. There are 15 contributors writing on isolation and maintenance of callus and cell suspension cultures, haploid cell cultures, plant protoplast methods, selection of plant cells, embryogenesis, organogenesis, plant regeneration, vascular differentiation, secondary product formation, cryopreservation and storage of germplasm, and tissue culture methods for virus and fungi.

363. **Plant Molecular Biology**. Edited by A. Weissbach and H. Weissbach. Orlando, Fla.: Academic, 1986. 829p. $83.00. ISBN 0121820181. (Methods in Enzymology, Vol. 118).

Volume 118 of this prestigious methods review series discusses plant molecular biology with focus on cell walls, membranes, the nucleus, chloroplast, transcription-translation, photosynthetic systems, mitochondria, nitrogen metabolism, gene transfer, and viralogy.

364. **Plant Tissue Culture 1982: Proceedings of the Fifth International Congress of Plant Tissue and Cell Culture**. Edited by Akio Fujiwara. Tokyo: Japanese Association for Plant Tissue Culture, 1982. 839p.

This proceedings from the Congress held at Tokyo and Lake Yamanaka, Japan, July 11-16, 1982, is a compilation of papers on recent progress in plant cell and tissue culture pertaining to both applied and basic research. It is appropriate for researchers and students in botanical and academic libraries. It may be updated by both *Cell Culture and Somatic Cell Genetics of Plants* (entry 334) and *Handbook of Plant Cell Culture* (entry 348).

365. Robertson, Kenneth R. **Observing, Photographing, and Collecting Plants**. Urbana, Ill.: Natural History Survey Division, State of Illinois, 1980. 62p. (Illinois Natural History Survey Circular 55).

This free booklet proposes to introduce the study of plants through observation, photography, and collection. It is clearly written for the amateur, and besides describing plant classification, identification, and structure, it discusses plant collection and photography in relation to equipment, materials, techniques, and herbarium specimens.

366. Sharma, Arun Kumar, and Archana Sharma. **Chromosome Techniques: Theory and Practice**. 3rd ed. London: Butterworths; distr., Stoneham, Mass.: Butterworths, 1980. 711p. $165.00. ISBN 0408709421.

This well-documented book includes procedures for both plant and animal systems using various kinds of physical and biochemical techniques.

367. **Tasks for Vegetation Sciences**. Vol. 1- . The Hague: Junk; distr., Hingham, Mass.: Kluwer Academic, 1981- . irregular. price varies. (Vol. 1: 1981. **Macroclimate and Plant Forms: An Introduction to Predictive Modeling in Phytogeography**. By Elgene Owen Box. 258p. ISBN 9061939410).

This series fuses mathematical concepts and analyses with techniques, methods, and instrumentation. Each volume serves as both a review of the literature and an exposition of new information.

368. **Techniques in Photomorphogenesis**. Edited by Harry Smith and Martin Geoffrey Holmes. Orlando, Fla.: Academic, 1984. 308p. ill. index. $65.00. ISBN 0126529906. (Biological Techniques Series).

Thirteen contributors, organized by these well-respected editors, intend this methodological and technological work to be used as a lab manual, with practical advice taking precedence over theory. It is included here despite its lab manual status because it is an authoritative guide for carrying out experiments on control of plant development by light. Contents include an introduction to the subject, criteria for photoreceptor involvement, light sources, techniques of radiation measurement, action spectroscopy, in vivo spectrophotometry, phytochrome purification and immunochemistry, information on blue light photoreceptors, and phytochrome in membranes. An appendix lists equipment sources. There is a wealth of information in this manual that can serve as a good survey or introduction to these techniques. As usual with a guide of this kind, though, it needs updating almost as soon as it comes off the press.

369. West, K. **How to Draw Plants: The Techniques of Botanical Illustration**. London: The Herbert Press in association with the British Museum (Natural History), 1983. 152p. ill. $22.50. ISBN 0823023559.

This book examines in detail the technical aspects of botanical illustrations. It includes a brief outline on the evolution of botanical illustration, with chapters on basic equipment, concepts, plant handling, plants in detail, and the use of various mediums. The reviewer called this book "a standard reference work for years to come." [R: KewMag, May 1984, pp. 93-94]

370. Wetherell, D. F. **Introduction to In Vitro Propagation**. Wayne, N.J.: Avery Publishing Group, 1982. 87p. $7.95pa. (Avery's Plant Tissue Culture Series).

Suitable for amateurs, this concise introduction explores the application of methods for specific interests by describing facilities, equipment, procedures, and elaboration of culture stages. Included are a glossary, list of suppliers, bibliography, and index. The book is especially recommended for high school students or hobbyists and as a useful addition to plant tissue culture laboratories. [R: QRB, Mar. 1983, p. 99]

371. Womersley, J. S. **Plant Collecting and Herbarium Development: A Manual**. Rome: Food and Agriculture Organization of the United Nations; distr., New York: Unipub, 1981. 148p. $10.75pa. ISBN 9251011443. (FAO Plant Production and Protection Paper, No. 33).

Womersley discusses practical information for the beginner, as well as specialized techniques for the more experienced botanist, with chapters on field collecting, preservation, identification, function and organization of the herbarium, processing of herbarium collections, herbarium curation, and ancillary services.

372.   Zweifel, Frances W. **A Handbook of Biological Illustration**. Chicago: University of Chicago, 1961. 131p. bibliog. $6.00pa. ISBN 0226996999. (Phoenix Science Series, PSS510).

This book discusses the materials and techniques of biological illustration, with emphasis on black-and-white drawing, color illustration, and photography.

6 Directories and Groups

Directories are alphabetical lists of names and addresses of people, organizations, manufacturers, products, periodicals, and the like, that present an array of information identifying or locating any chosen subject. Directories are often referred to as *tertiary* literature sources: they are *about* science rather than *of* science. Other examples of tertiary sources are guides to the literature, biographies, textbooks, and histories. Like these other sources, directories contain useful information presented as a convenient compilation of facts that may include indexes, supplementary text, or other information such as geographical location. They can be issued as part of a volume or periodical only once or as several volumes over a period of time. Directories in this chapter will be organized under the headings of biographical directories, commercial and government agency directories, noteworthy collections and Ph.D.-granting institutions, general directories of organizations, and lists of professional associations, learned societies, and research centers. These last two sections will lead the searcher to more information about specific societies.

Biographical Directories

Biographical directories can vary in content from simple lists of addresses to more complete information about a person's education, professional, and personal life. Membership lists can be considered directories, and many of the sources discussed in chapter 8, "Historical Materials," may contain just the sort of direction needed to satisfy the search for elusive details about an early scientist's life or an explorer's work. Following are some other examples of biographical directories.

373. **American Men and Women of Science: Physical and Biological Sciences**. 15th ed. New York: Bowker, 1982. 7v. $85.00/vol.; $495.00/set. ISBN 0835214133.

An alphabetical name arrangement provides biographical, educational, professional, research, and address information for 130,000 scientists in the physical and biological sciences. This directory is computerized and available online (AMWS). [R: LJ, Mar. 1, 1980, p. 602]

374. Botanical Society of America. **Directory**. Bloomington, Ind.: Department of Biology, Indiana University, 1986. 98p. $10.00. [Write to David L. Dilcher].

This directory includes names, addresses, and research specializations for current members of the Botanical Society. A list of officers and committees is provided. The directory is updated periodically.

375. **International Association of Bryologists Compendium of Bryology**.

See "Noteworthy Collections" section for information. This compendium includes a directory of bryologists.

376. **International Register of Specialists and Current Research in Plant Systematics**. Compiled and edited by Robert W. Kiger, T. D. Jacobsen, and Roberta M. Lilly. Pittsburgh, Pa.: Hunt Institute for Botanical Documentation, Carnegie-Mellon University, 1981. 346p. $10.00pa. ISBN 0913196398.

International in scope, this directory lists the names of over 1,500 scientists alphabetically by surname, including information on title, institution, position, address, telephone number, projects, and specialities. There are taxonomic, geographic, geologic, and methodology/general subject indexes to the main list of names. The *Register* is an ongoing project at the Hunt Institute and is available in printed and computerized form.

377. Knobloch, Irving William. **A Preliminary Verified List of Plant Collectors in Mexico**. Plainfield, N.J.: Moldenke, 1983. 179p. (Phytologia Memoirs, VI).

This is a list of all the main, and many of the casual, plant collectors in Mexico with reference to where their work was published or cited. There is a 64-page section listing references cited and an alphabetical listing of probable, but not verifiable, collectors.

Commercial and Government Agency Directories

378. **Annual Register of Grant Support: A Directory of Funding Sources**. Vol. 1- . Chicago: Marquis, 1967- . annual. $67.50/yr. ISSN 0066-4049.

This publication provides data on grant programs, money allotted, conditions and requirements for applications for funding from public, private, corporate, and non-traditional sources, such as educational, professional, church, or community agencies. Subject, organization, personnel, and geographic indexes are included.

379. **Computer-Readable Databases: A Directory and Data Sourcebook**. Martha E. Williams, ed.-in-chief. Chicago: American Library Association, 1984. 2v. $157.50/ set(pa.). ISBN 0838904157.

This comprehensive directory covers internationally available numeric and non-numeric databases in business, law, humanities, social sciences, science, technology, and medicine. Approximately 2,000 databases are described including data on name(s), update frequency, producer, subject, scope, data elements, database services, and user aids. There are subject, producer, vendor, and name indexes. Heavily endorsed by industry and professional leaders, *Computer-Readable Databases* is recommended for public and academic library collections. It is available online as DATABASE OF DATABASES. [R: LJ, Aug. 1985, pp. 64-65]

380. Coombs, Jim. **The Biotechnology Directory 1985: Products, Companies, Research and Organizations**. New York: Nature Press/Grove's Dictionaries of Music, 1985. 464p. index. $130.00pa. ISBN 0943818060.

This prize-winning directory covers more than 20 countries, over 4,000 organizations, and 950 products in the rapidly expanding area of biotechnology. Part 1 presents an introduction to biotechnology with an overview of current industrial applications, a directory of international biotechnological organizations, and a listing of information services, databases, periodicals, and newsletters. Part 2 provides national profiles of government agencies, societies, and associations actively involved in each of the countries selected for inclusion. Part 3 lists products and areas of research of noncommercial organizations, commercial companies, universities, and research centers. This comprehensive directory is exceptionally useful and recommended for a wide group of users and libraries.

381. **Data Base Directory**. Vol. 1- . White Plains, N.Y.: Knowledge Industry Publications, 1984- . monthly. $215.00/yr. ISSN 0749-6680.

This comprehensive service is published in cooperation with American Society for Information Sciences. In addition to the monthly issues, subscribers receive two directories per year, semiannual cumulative indexes of *DataBase Alert*, and a toll-free hotline access to the editors. Information includes access costs, producer contacts, producer addresses, telecommunication networks, annotation, producer services, language, copyright restrictions, etc. [R: LJ, Aug 1985, p. 65]

382. **Directory of Online Databases**. Vol. 1- . Santa Monica, Calif.: Cuadra Associates, 1979- . quarterly. $95.00/yr. ISSN 0193-6840.

This is another very useful comprehensive directory for online databases. Updated quarterly, it provides complete bibliographic, financial, and vendor information for over 3,000 online databases available through 450 online services worldwide.

383. **Foundation Directory**. Vol. 1- . New York: Foundation Center, 1960- . annual. $60.00/yr. ISSN 0071-8092.

This directory is available in print or online (FOUNDATIONS). The directory provides descriptions of more than 2,500 foundations, which have assets of $1 million or more, or which make grants of $500,000 or more annually. Each foundation is a "nongovernmental, nonprofit organization, with funds and program managed by its own trustees or directors, and established to maintain or aid social, educational, charitable, religious, or other activities serving the common welfare, primarily through the making of grants." Grants are awarded in the fields of education, health, welfare, sciences, international activities, and religion. The computerized database, covering data for one year, is updated and revised semiannually.

384. **Government Research Directory**. 3rd ed. Detroit: Gale, 1985. 675p. $325.00. ISBN 0810304635.

Formerly *Government Research Centers Directory*, this directory identifies and describes over 2,000 research and development organizations operated by the U.S. government or operated by others for, or in cooperation with, federal agencies. Entries cover agriculture, life sciences, medicine and health, business and commerce, conservation, education, energy, environment, science and technology, grants and contracts offices, and research administration offices. There are name, key word, and geographic indexes.

385. **The Software Catalog: Science and Engineering**. 3rd ed. New York: Elsevier Scientific Publishing, 1986. 625p. $49.50pa. ISBN 0444010580.

This subset of information is published in the *Software Catalog: Microcomputers* and the *Software Catalog: Minicomputers*. Produced from the International Software Database, Ft. Collins, Colorado, it includes 3,600 programs for all major micros and minis. Detailed descriptions of software are arranged by vendor and ISPN including information on vendor, address, complete bibliographic description, and an annotation. Indexes include computer system, operating system, programming language, microprocessor, subject and applications, key word and program name. This catalog includes a glossary and usage and ordering information. An example of one pertinent program is BOTANY FRUIT KEY (Sunnyvale, Calif.: Atari, $19.95, ISPN 27050-033). This is a new approach by Atari to identifying common fruits.

386. **Ulrich's International Periodicals Directory**. Vol. 1- . New York: Bowker, 1932- . annual. $110.00/yr. ISSN 0000-0175.

This classified guide to current domestic and foreign periodicals lists over 65,000 periodicals of all kinds from all over the world in almost 600 subject categories. Information for each entry includes title, frequency of publication, publisher name and address, country of publication code, Dewey Decimal Classification number, bibliographic and buying information, year first published, language, if advertising and book reviews are included, where abstracted or indexed, corporate author, and variant forms and names. It is available in printed form or as a computerized database that includes, in addition to the *International Periodicals Directory*, *Ulrich's Quarterly*, *Irregular Serials and Annuals*, and *Sources of Serials*.

Noteworthy Collections
and Ph.D-granting Institutions

Catalogs of other important collections may be found in chapter 1 with the bibliographic tools.

387. Botanical Society of America. **Guide to Graduate Study in Botany for the United States and Canada**. Columbus, Ohio: Botanical Society of America, 1983. 84p. $5.00.

This guide lists 82 plant science departments in the United States and 11 in Canada which offer the Ph.D. degree in some area of the plant sciences. Each departmental listing includes the name and address of the institution, name of the department with number of faculty, current graduate enrollment, fields of specialization represented in the department, and name, academic background, area of specialization, and titles of recent Ph.D. theses directed by members of the botanical faculty in the department. This publication is updated approximately every five years.

388. **Compendium of Bryology: A World Listing of Herbaria, Bryologists, and Current Research**. Compiled by Dale H. Vitt for the International Association of Bryologists. Forestburgh, N.Y.: Lubrecht and Cramer, 1985. 355p. $27.00. ISBN 3768214346. (Bryophytorum Bibliotheca, No. 30).

This world listing of herbaria, collectors, bryologists, and current research provides information on 535 current researchers in bryology, 471 herbaria, and location of 2,200 bryophyte collectors.

389. **Index Herbariorum**. Edited by Frans A. Stafleu. Utrecht, Netherlands: Bohn, Scheltema and Holkema. Part I: **The Herbaria of the World**, 7th ed. 1981. 452p. $69.00. ISBN 9031304786. (Regnum Vegetabile, Vol. 106).

This directory is a guide to the location and contents of the world's public herbaria providing a list of herbaria, with address, telephone number, sponsoring organization, collection strengths, names of director and staff, and services. It is arranged alphabetically by city name. There are indexes to important collections, geographical locations, and personal names. Between editions, the information is updated in *Taxon* (see entry 193).

390. **International Directory of Botanical Gardens**. 3rd ed. Compiled by D. M. Henderson and H. T. Prentice. Utrecht, Netherlands: Bohn, Scheltema and Holkema, 1977. 270p. $25.25. ISBN 9031302457. (Regnum Vegetabile, Vol. 95).

Arranged by country, this directory lists gardens by city with address, status, area, latitude and longitude, altitude, rainfall, taxa included, specialities, greenhouses, publications, accessibility, and names of the director and other staff. There are indexes to personal names as well as cities and gardens.

391. **Peterson's Annual Guides to Graduate Study. Book 3: Biological, Agricultural, and Health Sciences**. Vol. 1- . Princeton, N.J.: Peterson's Guides, 1966/67- . annual. $28.95/yr. ISSN 0278-5358.

This section of the annual directory to graduate study covers the botanical sciences. It is arranged by subject with an institution name index. Entries provide an abstract of the program of study for the institution and the department; information on research facilities, financial aid, cost of study, cost of living, student population, the area, the university, how to apply, address for additional information, and list of faculty with their credentials. Because this is updated annually, it provides complementary, current coverage to the more specialized directory published by the Botanical Society of America. [R: ARBA, 1977, entry 623]

392. **Preliminary Directory of Living Plant Collections of North America**. Edited by Charles A. Huckins. Prepared by the Plant Collections Committee of the American Association of Botanical Gardens and Arboreta, Inc. Swarthmore, Pa.: Association of Botanical Gardens and Arboreta, 1983. 73p.

This directory identifies important documented collections of living plants and other significant plant resources on the North American continent. It provides standard information on composition, location, origins, purposes, accessibility, management, utilization, and staffing. To be included in the directory, the collection must be open to the public or accessible for intended purposes. The intent is to update and revise the directory on a regular basis. Arranged alphabetically by location, the directory has indexes to plants, plant collections, 58 institutions, secondary locations, and natural or cultural heritage sites.

393.   **World Directory of Collections of Cultures of Microorganisms**. 2nd ed. Edited by Vicki F. McGowan and V. B. D. Skerman. Issued by World Data Center on Microorganisms with the support of UNESCO, FAO, WHO, UNU, UNIDO, UNEP, and the European Economic Commission. Queensland, Australia: World Data Center, 1982. 641p.

Five hundred and sixty-six collections are described with pertinent information on collections held, sponsors, address, staff, cultures, and services. There is explanatory material in English, French, German, Spanish, Russian, and Japanese. In addition to bacteria and protozoa, this directory also includes algae, fungi, yeasts, lichens, and plant viruses.

General Directories

394.   **Directory of Natural History and Related Societies in Britain and Ireland**. Edited by Audrey Meenan. London: British Museum (Natural History); distr., Charlottesville, Va.: University Press of Virginia, 1983. 407p. $25.00. ISBN 0565008595. (Publication No. 859).

This directory contains 750 entries covering all aspects of natural history, all relevant national societies and associations, local societies and groups, and university and school societies. The arrangement is alphabetical by society name with information on address, aims, publications, affiliations, membership, meetings, founding date, etc. There are geographical and subject indexes.

395.   **Encyclopedia of Associations, 1987**. 21st ed. Detroit: Gale, 1986. Vol. 1: **National Organizations of the U.S.** 3pts. $220.00/set. Vol. 2: **Geographic and Executive Index**. 799p. $200.00. Vol. 3: **New Associations and Projects**. 1v. Inter-edition subscription, $210.00. Vol. 4: **International Organizations, 1987**. ca. 800p. $185.00. ISSN 0071-0202.

The format provides multiple approaches to association services in the United States and abroad, covering nonprofit national organizations, new associations and projects, international organizations, research activities, and funding programs, with geographic and executive indexes. Information is given for organization name, acronym, key word, address, telephone number, chief official and title, founding date, number of members, staff, state and local groups, description, committees, sections and divisions, publications, affiliated organizations, mergers and name changes, and conventions/meetings. This is available as an online computerized database corresponding to the current edition of the encyclopedia.

396.   **European Sources of Scientific and Technical Information**. 6th ed. Detroit: Gale, 1984. 440p. $150.00. ISBN 0582901529.

Twelve hundred entries provide information on official name, address, telephone number, telex number, cable address, affiliation, contact person, subject coverage, and publications. This guide to key sources is arranged by subject and subdivided by country, with title, key word, and subject indexes.

397. **International Research Centers Directory**. 2nd ed. Detroit: Gale, 1984. 739p. $310.00. ISBN 0810304678.

Included in this directory are 1,500 international research organizations to cover university-related, independent, and government research organizations in a wide range of subject areas. This source is complementary to *Research Centers Directory* (entry 398) and *Government Research Directory* (entry 384).

398. **Research Centers Directory**. 10th ed. Detroit: Gale, 1985. 2v. $340.00/set. ISBN 0810304694.

This comprehensive guide to university-related and other nonprofit research centers in the United States and Canada includes 8,200 entries covering institutes, centers, foundations, laboratories, bureaus, experiment stations, and research support units. There are institutional, subject, acronym, alphabetical research center, and special capabilities indexes.

399. **World Guide to Scientific Associations and Learned Societies**. 4th ed. New York: K. G. Saur, 1984. 947p. $112.00. ISBN 3598205228.

Twenty-two thousand organizations in all fields are covered in this one-volume guide. Scope is international; the strength of this directory lies in its breadth. There are indexes by association abbreviations and subject. [R: Nature, Dec. 13, 1984, pp. 668-69]

400. **World of Learning 1986**. 36th ed. Detroit: Gale, 1986. 1896p. $170.00. ISBN 0946653135.

This outstanding publication serves as a guide to the world's universities, colleges, libraries, museums, and learned societies. Over 24,000 entries, 15,000 professors, and 400 cultural organizations are included. This is another source of international information about botanical institutes, museums, and libraries.

Professional Associations,
Learned Societies, and Research Centers

Following is a list of professional organizations and societies, with their addresses and primary activities, that may be of interest to botanical hobbyists or researchers. Although the scope is international, emphasis is on U.S. associations with either specialized, or broad, botanical interests.

401. **American Association of Botanical Gardens and Arboreta**. P.O. Box 206, Swarthmore, PA 19081.

The group consists of directors, staffs of botanical gardens, arboreta, and institutions maintaining or conducting courses, etc. Its primary function is to bestow awards and sponsor placement service. Publications include a newsletter, bulletin, and proceedings. There is an annual meeting.

402. **American Bryological and Lichenological Society**. c/o Barbara Crandall-Stotler, Botany Department, Southern Illinois University, Carbondale, IL 62901.

Professional botanists, botany teachers, and hobbyists interested in the study of mosses, liverworts, and lichens constitute this Society, which publishes *Bryologist*. There is an annual meeting.

403. **American Fern Society**. Department of Botany, University of Tennessee, Knoxville, TN 37996.

This group was founded to promote the study of ferns and their allies. Publications include a newsletter and the *American Fern Journal*. An annual meeting is held in conjunction with the American Institute of Biological Sciences.

404. **American Institute of Biological Sciences**. 1401 Wilson Blvd., Arlington, VA 22209.

This Institute consists of professional biological associations and industrial firms whose members have an interest in the life sciences. Its aims are to promote unity and effectiveness of effort among persons engaged in biological research, teaching, and application of biological data and to further relationships of biological sciences with other sciences, the arts, and industries. AIBS conducts symposia, arranges speakers' services, provides consultants and advisory committees to government agencies, and publishes *Bioscience*. There is an annual meeting.

405. **American Magnolia Society**. P.O. Box 129, Nanuet, NY 10954.

Botanists, horticulturists, commercial growers, and people interested in the study of the magnolia family are welcome to join. The group conducts international registration of magnolia clones and parentage of hybrids. Publications include *The Magnolia Journal* and *Magnolia Checklist* done in cooperation with the American Horticultural Society. There is an annual convention.

406. **American Orchid Society**. 84 Sherman St., Cambridge, MA 02140.

This Society consists of professional growers, botanists, and hobbyists who are interested in promoting all phases of orchidology. It sponsors the World Orchid Conference and bestows 1,000 awards each year. It publishes a bulletin and an awards quarterly. There is an annual fall meeting.

407. **American Phytopathological Society**. 3340 Pilot Knob Rd., St. Paul, MN 55121.

Researchers, educators, and others interested in the study and control of plant disease constitute this group. Publications are *Phytopathology*, *Plant Disease*, *New Fungicide and Nematocide Data*, a directory of members, series of classic reprints, monographs, disease compendia, etc.; teaching films are also produced. There is an annual meeting.

408. **American Plant Life Society**. P.O. Box 985, National City, CA 92050.

The group was organized to study the plant family Amaryllidaceae and various related families. Publications are *American Amaryllid*, *Herbertia*, and *Amaryllis Handbook*.

409. **American Society of Plant Physiologists.** P.O. Box 1688, Rockville, MD 20850.
This is a professional society of plant physiologists and plant biochemists engaged in research and teaching. The Society bestows awards and offers a placement service. Publications include *Plant Physiology*, a newsletter, directory of members, and program and abstracts of the annual meeting.

410. **American Society of Plant Taxonomists.** c/o Dr. Neil Harriman, Biology Department, University of Wisconsin, Oshkosh, WI 54901.
Members are botanists and others interested in all phases of plant taxonomy. Publications include *Systematic Botany*, a membership directory, and the series, Systematic Botany Monographs. An annual meeting usually is held in conjunction with the American Institute of Biological Sciences.

411. **Association for Tropical Biology.** c/o Smithsonian Institution, Washington, DC 20560.
This is an international association of people interested in tropical biology who seek to coordinate existing knowledge and provide new knowledge about the plants and animals of the tropics. Publications include *Biotropica* and irregular symposia proceedings. The annual meeting is held in conjunction with the American Institute of Biological Sciences.

412. **Association of Systematics Collections.** Museum of Natural History, University of Kansas, Lawrence, KS 66045.
The purpose of this organization is to foster the care, management, preservation, and improvement of systematics collections and to facilitate their utilization in science and society by the following means: providing representation for institutions housing systematics collections; encouraging direct interaction among those concerned with systematics collections; providing a forum for consideration of mutual problems; and promoting the role of systematics collections in research, education, and public service. Publications include the newsletter, museum collections and computers, sources of federal funding, and selected monographs. They also maintain a computerized registry for taxonomic resources and services.

413. **Botanical Society of America.** c/o Carol C. Baskin, School of Biological Sciences, University of Kentucky, Lexington, KY 40506.
This group consists of professional botanists and others engaged in plant science. It conducts special research programs and sponsors summer institutes for college teachers. Publications include *American Journal of Botany*, *Plant Science Bulletin*, *A Guide to Graduate Study in Botany for the U.S. and Canada*, a career bulletin, and membership directory. The annual meeting usually is held in conjunction with the American Institute of Biological Sciences.

414. **Botanical Society of the British Isles.** c/o Department of Botany, British Museum (Natural History), Cromwell Rd., London SW7 5BD.
This group studies British flowering plants and ferns, arranges conferences, and publishes conference reports and a newsletter. The eighteenth conference report was published as *Plant-Lore Studies: Papers Read at a Joint Conference of the Botanical Society of the British Isles and the Folklore Society, Held at the University of Sussex, April, 1983*, edited by Roy Vickery (London: The Folklore Society, 1984, 261p., ISBN 0903515083).

415. **Canadian Botanical Association/L'Association botanique du Canada**. Department of Biology, Carleton University, Ottawa, ON K1S 5B6 Canada.
This group publishes the *CBA/ABC Bulletin*.

416. **Canadian Phytopathological Society/La Société canadienne de phytopathologie**. Plant Science Department, University of British Columbia, Vancouver, BC V6T 1W5 Canada.
Canadian Journal of Plant Pathology and *CPS News* are the Society's publications.

417. **Canadian Society of Plant Physiologists/La Société canadienne de physiologie vegetale**. Department of Biology, University of Calgary, 2500 University Dr. N.W., Calgary, ALTA T2N 1N4 Canada. The Society publishes the *C.S.P.P. Newsletter*.

418. **Council of Biology Editors**. 9650 Rockville Pike, Bethesda, MD 20814.
This group of active and former editors of primary and secondary journals in the biosciences conducts study and discussion groups, panels, and workshops to consider all aspects of bioscience communication with emphasis on publication, especially in primary journals and retrieval in secondary media. Publications are *CBE Views, Style Manual, Economics of Scientific Journals*, and *Scientific Writing for Graduate Students*. There is an annual meeting.

419. **Council on Botanical and Horticultural Libraries (CBHL)**. Treasurer: John F. Reed, The New York Botanical Garden, Bronx, NY 10458-5126.
Membership is open to institutions and to individuals. The purpose of CBHL is to initiate and improve communication between persons and institutions concerned with the development, maintenance, and use of libraries of botanical and horticultural literature. CBHL encourages and assists in the coordination of activities and programs of mutual interest and benefit. The group publishes a quarterly newsletter and a plant bibliography series. There is an annual spring meeting.

420. **Ecological Society of America**. Department of Biology, Notre Dame University, Notre Dame, IN 46556.
Educators, professional ecologists, and scientists interested in the study of plants, animals, and man in relation to their environment have formed this group, which seeks to develop better understanding of biological processes and their contribution to agriculture, wildlife and range management, industries, public health, and conservation. Publications include *Ecology, Ecological Monographs*, membership directory, and the *Bulletin*. The Society, affiliated with the American Institute of Biological Sciences, holds an annual meeting.

421. **Herb Research Foundation**. 1780 55th St., Boulder, CO 80301.
Membership includes professionals, practitioners, and researchers in the health food industry, ethnobotanists, and interested consumers. The group encourages and supports research on the chemistry and pharmacology of herbal folk medicines, teas, and other botanical products; provides a forum for discussion and cooperation; forms a liaison between the U.S. herbal movement and the worldwide scientific community; and publishes and disseminates research information on botanical products. There is an annual meeting.

422. **Herb Society of America**. 191 Sudbury Rd., Concord, MA 01742.

Scientists, educators, and others interested in botanical and horticultural research on herbs and the culinary, economic, decorative, fragrant, and historic use of herbs constitute this Society. The group maintains herb gardens, a speakers' bureau, and bestows awards. Publications include a newsletter, a membership directory, and the *Herbalist*. There is an annual convention.

423. **International Association for Plant Physiologists**. Secretary/Treasurer: Dr. D. Graham, Plant Physiology Group, Food Research Lab CSIRO, P.O. Box 52, North Ryde, NSW 2113 Australia.

This group promotes the development of plant physiology at the international level. Publications include a newsletter, and *Tentative Recommendations of Terminology, Symbols and Units in Plant Physiology*.

424. **International Association for Plant Taxonomy**. Secretary: Frans A. Stafleu, Room 1902, Transitorium II, Heidelberglaan 2, 3508 TC, Utrecht, Netherlands.

The aims of this group are to promote the development of plant taxonomy and to encourage international relations between interested individuals and institutes. Publications include *Taxon*, the Regnum Vegetabile series, manuals, and reference works of various sorts.

425. **International Association for Plant Tissue Culture**. Headquarters: c/o Lab of Plant Nutrition, Department of Agricultural Chemistry, Faculty of Agriculture, Kyoto University, Yoshida-Hommachi/Sakyo-ku, Kyoto 606 Japan.

The fifth congress was held in 1982.

426. **International Association of Botanic Gardens**. Secretary: Dr. Roy L. Taylor, Botanical Garden, University of British Columbia, 6501 NW Marine Dr., Vancouver, BC V6T 1W5 Canada.

This Association aims to promote international cooperation between botanic gardens, arboreta, and other similar institutes maintaining scientific collections of living plants. Meetings are held every six years at the International Botanical Congresses and every four years at the International Horticultural Congresses.

427. **International Botanical Congress**. Executive Secretary: W. J. Cram, Australian Academy of Science, P.O. Box 783, Canberra ACT 2601 Australia.

The aims of this Congress are the consideration and discussion of recent results of fundamental botanical studies and scientific aspects of practical application and the development of personal contacts among botanists. Congresses, which are open to all botanists, alternate between Europe and other locations. The next Congress will be held in Berlin in 1987. Publications include Congress proceedings and abstracts.

428. **International Council of Botanic Medicine**. President: Dr. Jacob E. Thuna, 11 St. Catherine St. East, Montreal 129 PQ Canada.

The Council educates in the science of botanic medicine and cooperates with medical and naturopathic herbalists' societies, as well as professional schools, on the ethical practice of botanic medicine. Publications include *Journal of Natural Therapeutics, Health from Herbs*, and *The Herbal Practitioner*.

429. **International Organization of Plant Biosystematists**. President: Professor William F. Grant, Box 4000, Genetics Laboratory, MacDonald Campus of McGill University, St. Anne de Bellevue, PQ H9X 1CO Canada.

This group is affiliated with the International Association for Plant Taxonomy and holds meetings at the International Botanical Congress. It also publishes a newsletter.

430. **International Photosynthesis Committee**. Professor D. O. Hall, Kings College, University of London, 68 Half Moon Lane, London SE24 9JF England or c/o Dr. J. Biggins, Division of Biology, Brown University, Providence SE24, RI 02912.

Founded in 1968, this Committee organizes the triennial International Photosynthesis Congress, last held in 1986.

431. **International Phycological Society**. Secretary: Dr. Sylvia A. Earle, California Academy of Sciences, Golden Gate Park, San Francisco, CA 94118.

This Society aims to promote development of phycology, facilitate dissemination of phycological information about algae, and bring about international cooperation among phycologists. It publishes *Phycologia*.

432. **International Society for Medicinal Plant Research**. Contact: Professor Dr. O. Sticher, Eidg Technische Hochschule, Pharmazeutisches Institut, CH-8092, Zurich, Switzerland.

The last meeting was held in Zurich in 1978.

433. **International Society of Plant Morphologists**. Secretary/Treasurer: Professor H. Y. Mohan Ram, Department of Botany, University of Delhi, Delhi 110007 India.

Publications include *Phytomorphology*, *Plant Tissue and Organ Culture: A Symposium*, and *Recent Advances in the Embryology of Angiosperms*.

434. **Mycological Society of America**. c/o Harold H. Burdsall, Jr., P.O. Box 5130, Center for Forest Mycology Research, Forest Products Lab, Madison, WI 53705.

Members include professional mycologists and others interested in the study of fungi through research, teaching, and industrial application. Publications include *Mycologia*, a newsletter, and *Mycologia Memoirs*. There is an annual meeting.

435. **North American Mycological Association**. 4245 Redinger Rd., Portsmouth, OH 45662.

The Association promotes amateur mycology, sponsors field trips, taxonomy and mycological seminars, assists in classifying specimens, presents amateur mycological awards, and sponsors contests. Publications include *Mycophile* and *McIlvainea*. The group is affiliated with the Mycological Society of America.

436. **Organization for Flora Neotropica**. Executive Director: Dr. Ghillean T. Prance, New York Botanical Garden, Bronx, NY 10458.

This organization produces and publishes a general and complete flora of the tropical American region, using services and skills of specialized botanists throughout the world. Publications include a monograph series.

437. **Phycological Society of America**. c/o Carole Lembi, School of Life Sciences, University of Nebraska, Lincoln, NE 68588.

Educators, researchers, and others interested in the pure, applied, or avocational study and utilization of algae have formed this group. Publications are *Journal of Phycology*, a newsletter, and a membership directory. It is affiliated with the American Association for the Advancement of Science, with whom the annual meeting usually is held.

438. **Phytochemical Society of Europe**. Department of Biochemistry, University College of Swansea, Swansea SA2 8PP England.

Publications include the *Annual Proceedings* of the Society (ISSN 0309-9393), which provides a record of the annual meeting. Volume 26 (1985) is titled, *Plant Products and the New Technology*.

439. **Phytochemical Society of North America**. Biology Department, University of South Florida, Tampa, FL 33620.

This Society consists primarily of research scientists interested in all aspects of the chemistry of plants with the aim of promoting phytochemical research and communication. It publishes *Recent Advances in Phytochemistry*, which are the proceedings of the annual meeting. It also issues a newsletter and a membership directory.

440. **Society for Economic Botany**. Biological Sciences Group U-43, University of Connecticut, Storrs, CT 06268.

Botanists, anthropologists, pharmacologists, and others interested in scientific studies of useful plants may join this group. It hosts conferences, seeks to develop interdisciplinary channels of communication, and presents awards to outstanding economic botanists. Publications include *Economic Botany*, a membership directory, and proceedings of occasional symposia. This Society holds an annual meeting and is affiliated with the American Association for the Advancement of Science.

441. **Torrey Botanical Club**. c/o New York Botanical Garden, Bronx, NY 10458.

This long-standing Club, believed to be the first botanical society founded in America, consists of botanists and others interested in botany and in collecting and disseminating information on all phases of plant science. It has an international membership, although active membership is concentrated in the New York City area. The group sponsors field trips, publishes original investigations in all phases of botany, and presents an annual award for the best paper written by a student and published in the *Torrey Botanical Club Bulletin*. Publications include the *Bulletin* and *Memoirs of the Torrey Botanical Club*. Meetings are held biweekly, from October through December, and from March through May. The Club is affiliated with the American Institute of Biological Sciences.

7 Identification Sources
Atlases, Field Guides, Floras, Keys, and Manuals

This chapter annotates sources useful for identifying and locating plants. References in chapter 4, "Dictionaries and Encyclopedias," also may be helpful in this regard. Because there is a large number of very good field guides and manuals in print, this section includes a selected list of some of the best: standard sources that provide reliable, accurate information. Emphasis is on recent, in-print material of lasting value. Alternatives are included for more complete coverage; most libraries, except for the largest research collections, would not contain all the available sources on a particular topic. References for regional materials are rarely included here; instead, sources for more general works are listed, with reliance on the guides to floras listed below for specific information on particular regions or taxonomic divisions.

Guides to Floras

442. **Bibliographical Contributions from the Lloyd Library.** Vol. 1: **Bibliography of Floras of Europe, Great Britain, North America, South America, Asia, Africa.** Cincinnati, Ohio: Lloyd Library, 1911-18. For more information, see entry 24.

443.   Blake, Sidney Fay. **Guide to Popular Floras of the United States and Alaska**. Washington, D.C.: Government Printing Office, 1954. 56p. (U.S. Department of Agriculture Bibliographical Bulletin, 23).

This is an annotated, selected list of nontechnical works for the identification of flowers, ferns, and trees.

444.   Blake, Sidney Fay, and Alice Cary Atwood. **Geographical Guide to Floras of the World**. Part 1: **Africa, Australia, North America, South America and Islands of the Atlantic, Pacific, and Indian Oceans**. Washington, D.C.: Government Printing Office, 1942; repr., New York: Hafner, 1967. 336p. Part 2: **Western Europe, Finland, Sweden, Norway, Denmark, Iceland, Great Britain with Ireland, Netherlands, Belgium, Luxembourg, France, Spain, Portugal, Andorra, Monaco, Italy, San Marino, Switzerland**. Washington, D.C.: Government Printing Office, 1961; repr., Monticello, N.Y.: Lubrecht and Cramer, 1974. 742p. $56.00. ISBN 3874290603.

Updated and expanded by Frodin (entry 446).

445.   Boivin, Bernard. "A Basic Bibliography of Botanical Biography and a Proposal for a More Elaborate Bibliography." **Taxon** 26, no. 1 (1977): 75-105.

Although this article basically deals with biographical materials, it also has a section on regional collections of biographies that lists regional floras.

446.   Frodin, D. G. **Guide to Standard Floras of the World**. New York: Cambridge University Press, 1985. 619p. $175.00. ISBN 0521236886.

"An annotated, geographically arranged systematic bibliography of the principal floras, enumerations, checklists, and chorological atlases of different areas." The subtitle basically describes this "remarkable compendium." [R: Nature, May 2, 1985, p. 82]

447.   Harrington, Harold David. **How to Identify Grasses and Grasslike Plants (Sedges and Rushes)**. Chicago: Swallow Press, 1977. 142p. ill. $8.95pa. ISBN 0804007462.

This field guide includes a section on literature of the grasses.

448.   Lawrence, George H. M. **Taxonomy of Vascular Plants**. New York: Macmillan, 1951. 823p. contact publisher for price. ISBN 002368190X.

Older systematic literature is listed and annotated.

449.   Merrill, Elmer Drew, and Egbert H. Walker. **A Bibliography of Eastern Asiastic Botany**. Jamaica Plain, Mass.: Arnold Arboretum of Harvard University, 1938. 719p. **Supplement 1**. By Egbert H. Walker.

For more information, see entry 51.

450.   Radford, Albert E., et al. **Vascular Plant Systematics**. New York: Harper & Row, 1974. 891p. ill. $28.50. ISBN 0060453095.

There are several sections listing and describing floras and field guides.

451.   Reed, Clyde F. **Bibliography to Floras of Southeast Asia**. See entry 59 for information about Southeast Asian floras.

452. Sachet and Fosberg. **Island Bibliographies**. See entry 62 for information on island floras.

453. **Scientific and Technical Books and Serials in Print**. Vol. 1- . New York: Bowker, 1978- . irregular. $59.00/yr. ISSN 0000-054X.
Consult the subject section of this series for floras in print.

Atlases, Field Guides, Floras, Keys, and Manuals

454. Ammirati, Joseph F., James A. Traquair, and Paul A. Horgen. **Poisonous Mushrooms of the Northern United States and Canada**. Minneapolis, Minn.: University of Minnesota Press, 1985. 396p. ill. (col.). bibliog. index. $75.00. ISBN 0816614075.
This carefully researched and well-reviewed book contains excellent colored plates and line drawings, plus definitive descriptions and keys. It provides an overview of poisonous mushrooms that inhabit the northern United States with technical information on habitat, habit, occurrence, characteristics, and useful literature. It is aimed at researchers, physicians, and amateurs and will provide an outstanding, standard reference source. Besides discussing fungal structure and taxonomy, the book also deals with toxicity and the major categories of fungal poisoning.

455. Anderson, D. A. **All the Trees and Woody Plants of the Bible**. Waco, Tex.: Word Books, 1979. 294p.
This popular book for lay readers of the Bible discusses 76 trees and woody plants of the Bible with citations to the King James version. Related chapters describe tree gardens, shipbuilding, wood products, memorable trees, forest fires, and environmental aspects. This reference can be used conveniently with Walker (entry 551); although the books overlap in subject, Anderson includes some plants not in Walker. Walker's full-page watercolor illustrations, however, are superior to the mostly black-and-white photographs in Anderson. For a more comprehensive, scholarly treatment, see Moldenke (entry 526) or Zohary (entry 555). [R: ARBA, 1981, entry 1451]

456. Angier, Bradford. **Field Guide to Edible Wild Plants**. Harrisburg, Pa.: Stackpole, 1974. 256p. ill. $9.95pa. ISBN 0811720187.
"A quick all-in-color identifier of more than 100 edible wild foods growing free in the United States and Canada." There are alphabetically arranged descriptions of 116 free-growing, edible plants, all with illustrations in color. Information includes habitat, distribution, growth and development, edible parts, preparation and storage information, and general uses. The book is written by a well-known author of camping and outdoor living literature and its size is suitable for field work. It would be most useful in public libraries and for students.

457. Angier, Bradford. **Field Guide to Medicinal Wild Plants**. Harrisburg, Pa.: Stackpole, 1978. 320p. ISBN 0811720764.
This other book by Angier is a popular guide to interesting, valuable, and/or engrossing plants used for medicinal purposes. Arranged by common name, it gives family, other common names, characteristics, distribution, uses, and full-page color

illustrations for each plant. It is designed as a field guide, and like the *Field Guide to Edible Wild Plants* (entry 456), it is appropriate for the lay reader or amateur botanist.

458. **Atlas of the Flora of the Great Plains**. The Great Plains Flora Association. Edited by Theodore Mitchell Barkley. Ames, Iowa: Iowa State University Press, 1977. 600p. $26.00. ISBN 0813801354.

This definitive work is described by its editor as providing "distributional information for the vascular plants growing in the Great Plains of North America," and is a precursor for a *Flora of the Great Plains* (see entry 486) that will provide keys, descriptions, illustrations, and nomenclature for 3,000 taxa occurring in the area. The contents of the present atlas include a list of plant families, a distribution map showing county names, and a section of distribution maps for 2,217 taxa. There is an index of plant family names, generic names, and colloquial names.

459. **Audubon Society Field Guides.**

The Audubon Society guides are attractive, authoritative, pocket-size publications that are profusely illustrated. An outstanding value, they are highly recommended for school, academic, and public libraries. Following are examples of the wide range of guides available under the auspices of the Audubon Society:

Brown, Lauren. **Grasslands**. New York: Alfred A. Knopf, 1985. 606p. $14.95. ISBN 0394731212.

Little, Elbert L. **The Audubon Society Field Guide to North American Trees: Eastern Region**. New York: Alfred A. Knopf, 1980. 714p. $13.50. ISBN 0394507606.

Little, Elbert L. **The Audubon Society Field Guide to North American Trees: Western Region**. New York: Alfred A. Knopf, 1980. 639p. $12.00. ISBN 0394507614.

Niering, William A., and N. C. Olmstead. **The Audubon Society Field Guide to North American Wildflowers: Eastern Region**. New York: Alfred A. Knopf, 1979. 863p. $13.50. ISBN 0394504321.

Spellenberg, Richard. **The Audubon Society Field Guide to North American Wildflowers: Western Region**. New York: Alfred A. Knopf, 1979. 862p. ill. index. $13.50. ISBN 0394504313.

460. Barnett, Horace L., and Barry B. Hunter. **Illustrated Genera of Imperfect Fungi**. 3rd ed. Minneapolis, Minn.: Burgess, 1972. 241p. $13.95 (spiral-bound). ISBN 0808702661.

This identification manual "continues to be useful both in the scientific library and in the mycology laboratory" for work with a most difficult plant group. [R: ARBA, 1973, entry 1439]

461. Barnett, James Arthur, R. W. Payne, and D. Yarrow. **Yeasts: Characteristics and Identification**. New York: Cambridge University Press, 1984. 811p. ISBN 0521252962.

This huge reference provides commonly used procedures for yeast identification and gives detailed descriptions for 473 yeast species. There are 18 identification keys, 83 tests for identification, excellent photomicrographs, and a section containing specific epithets and yeast names. Replacing an earlier key and guide written by Barnett, et al., *Yeasts* is accurate, authoritative, and will "undoubtedly stand as the major text on yeast taxonomy for considerable time to come." The text has been revised as a software program available for $175.00 (New York: Cambridge University Press). [R: Nature, Jan. 26, 1984, p. 394]

462. Batson, Wade T. **Genera of the Western Plants: A Guide to the Genera of Native and Commonly Introduced Ferns and Seed Plants of North America, West of about the 98th Meridian and North of Mexico**. Columbia, S.C.: Published by the author, 1984. 209p. $8.95pa. ISBN 0872494519.

463. Batson, Wade T. **Guide to the Genera of the Plants of Eastern North America**. rev. 3rd ed. Columbia, S.C.: University of South Carolina Press, 1984. 203p. $8.95pa. ISBN 0872494500.

These field guides are compact publications describing the genera of the plants of North America. They provide keys for identification, drawings, and information on name authority, common names, synonyms, characteristics, habitat, and range.

464. Bianchini, Francesco, and Francesco Corbetta. **Health Plants of the World: Atlas of Medicinal Plants**. New York: Newsweek Books, 1977. 242p. ISBN 088225250X.

This beautiful book presents full-page color illustrations for each plant included. The plants are grouped by principal areas of operation, giving historical and chemical information, uses, distribution, etymology, and appropriate quoted remarks taken from literature. An appendix lists brief scientific information; there are botanical and pharmacological glossaries. A short bibliography and an index are also included. This book was written for plant lovers, not scholars, and is more suitable for public or school libraries than for academic libraries.

465. Bresadola, Giacomo. **Iconographia Mycologica**. Milan, Italy: Societa Botanica Italiana, Sezione Lombarda, 1927-60. 28v., with supplements.

This classic work, reprinted by the publisher in 1981, contains more than 1,500 beautifully drawn color plates, showing more than 1,650 species of fungi. The text is in Italian.

466. Brockman, Christian Frank. **Trees of North America: A Guide to Field Identification**. Racine, Wis.: Western Publishing Co., 1968. 280p. (A Golden Field Guide).

Nomenclature for this field guide is based on Little's *Checklist* (entry 559); this guide is useful in conjunction with the manuals and atlases by Little (entry 514) and Sargent (entry 541). It describes, with illustrations, 730 major species of trees native to North America north of Mexico. This compact guide is designed for use in the field.

467. Brouk, B. **Plants Consumed by Man**. New York: Academic, 1975. 479p. $73.50. ISBN 012136450X.

Over 300 plants used in various ways for human consumption are described, although there is no attempt to be exhaustive in this reasonably comprehensive reference. It is arranged by broad categories of plant products, then alphabetically by common name. Information provides etymology, history, geography, chemistry, morphology, physiology, and other items of interest. A glossary, bibliography, and an index are provided.

468. Brown, Lauren. **Weeds in Winter**. New York: Norton, 1976. 252p. ISBN 0393064115.

This introduction for the nonspecialist discusses how to identify dried plants, how to use the key, and where to go for additional information. The book, covering the flora of the East Coast of the United States, describes and illustrates winter aspects of over 135 common nonwoody wild plants. It will serve as a valuable aid on winter walks through the countryside. [R: ARBA, 1978, entry 1309]

469. **Common Weeds of the United States**. Prepared by the Agricultural Research Service of the U.S. Department of Agriculture. Washington, D.C.: U.S. Department of Agriculture, 1970; repr., New York: Dover, 1971. 463p. $8.95pa. ISBN 0486205045.

Originally titled *Selected Weeds of the United States*, the purpose of this valuable reference source is to assist with weed identification so that control can be established. Illustrations and nontechnical descriptions for many of the important U.S. weeds, covering 224 species, are included. There are 220 maps and 225 drawings in this helpful, attractive, and careful work by the well-known botanist, Clyde Franklin Reed. [R: ARBA, 1972, entry 1554; Phyto, May 1972, p. 317]

470. Coon, Nelson. **The Dictionary of Useful Plants**. Emmaus, Pa.: Rodale, 1974. 290p. bibliog. index. ISBN 087857185Xpa.

"The use, history, and folklore of more than 500 plant species." This bulk of this very useful book is arranged alphabetically by family with information on botanical and common names, synonyms, location, history, and uses of plants for food, medicine, and crafts. There is an extensive bibliography of more than 200 books and articles on plant uses. Appropriate for public libraries, this is one of the more comprehensive books on the subject.

471. Cooper, Marion R., and Anthony W. Johnson. **Poisonous Plants in Britain and Their Effect on Animals and Man**. London: Her Majesty's Stationery Office, 1984. 300p. ill. bibliog. index. £12.95. ISBN 0112425291. (MAFF Reference Book 161).

Written for people with a scientific background in botany, this reference provides comprehensive, practical information. Over 250 plants and 30 fungi are discussed with data on habitat, appearance, poisonous properties, incidence of poisoning, and suggestions for treatment.

472. Crockett, Lawrence J. **Wildly Successful Plants: A Handbook of North American Weeds**. New York: Macmillan, 1977. 268p. ISBN 0025288504; 0020626002pa.

This book provides excellent descriptions for over 100 common weeds, including information on habitat, control, and characteristics. There are full-page drawings and a simple identification system. Although Crockett cannot compete with the more comprehensive *Common Weeds of the United States* (entry 469), it does a good job for the species that it does cover. [R: ARBA, 1978, entry 1310; Choice, Oct. 1977, p. 1086]

473. Davis, Peter Hadland, and James Cullen. **The Identification of Flowering Plant Families, Including a Key to Those Native and Cultivated in North Temperate Regions**. 2nd ed. New York: Cambridge University Press, 1979. 113p. ill. $29.95; $8.95pa. ISBN 0521221110; 0521293596pa.

Introductory chapters discuss usage of terms and how to examine the plant for identification purposes. There are keys followed by brief descriptions of each family. The guide can be used with *Flowering Plants of the World* (entry 279) and is suitable for amateurs or professional botanists to identify all flowering plant families found wild or cultivated. [R: AAAS, Vol. 15, 1979-80, p. 212]

474.   Duke, James A. **CRC Handbook of Medicinal Herbs**. Boca Raton, Fla.: CRC Press, 1985. 677p. ill. index. $180.00. ISBN 0849336309.

This catalog discusses over 365 species of folk medicinal herbs whose safety is, or has been, in question by various federal agencies. Data include scientific and common names, descriptions of uses, medicinal applications, chemical content, and toxicity. Tables include information on (1) medicinal herbs: toxicity ranking and price list; (2) toxin distribution in the plant genera; (3) higher plant genera toxins; (4) pharmacologically active phylochemicals; and (5) analyses of conventional plant foods. There is an extensive list of references.

Another book of interest by the same author, who is Chief of the Germplasm Research Laboratory, U.S. Department of Agriculture, is *Medicinal Plants of the Bible* (Owerri, Nigeria: Trado-Medic, 1983, 300p., $49.95, ISBN 0932426239).

475.   Elias, Thomas S. **The Complete Trees of North America: Field Guide and Natural History**. New York: Van Nostrand Reinhold, 1980. 864p. $22.95. ISBN 0442238622.

The purpose of this guide is to help identify "quickly and confidently" any of over 750 North American trees. The reviewer in *ARBA* called this a rival to Sargent (entry 541), but a more useful book in many ways. Appropriate for the specialist and the amateur, it includes a diversity of information on trees of North America. The keys and descriptions are clear, useful, and easy to use. Refer to both Sargent (entry 541) and Little (entry 514) for alternatives. [R: ARBA, 1982, entry 1469]

476.   Elias, Thomas S., and Peter A. Dykeman. **Field Guide to North American Edible Wild Plants**. New York: Outdoor Life; distr., New York: Van Nostrand Reinhold, 1983. 286p. $19.95pa. ISBN 0442222548.

The descriptions, locations, identification, harvesting, and preparation of 200 of the most popular and common wild plants are contained in this beautifully illustrated and organized book. It is appropriate for public libraries and botany collections and is an alternative to Angier (entry 456), Brouk (entry 467), and Hall (entry 498).

477.   Elliott, Douglas B. **Roots: An Underground Botany and Forager's Guide**. Old Greenwich, Conn.: Chatham, 1976. 160p. $7.95pa. ISBN 0856991325.

"The useful wild roots, tubers, corms and rhizomes of North America." This book was written for the amateur and includes introductory material on plant names as well as the structure and function of roots. It is arranged according to habitat for roots of shade-loving forest plants, roots of aquatic plants, etc. There is an appendix on harvesting and examining root systems, a glossary, a bibliography, and an index. Descriptions provide scientific and common names, illustrations, habitat, distribution, characteristics, and uses.

478.   Everard, Barbara. **Wild Flowers of the World**. Paintings by Barbara Everard and text by Brian D. Morley. New York: Putnam, 1970. 432p.

Over 1,000 plants are illustrated in this selection of beautiful wildflowers from around the world, grouped geographically. There is an introduction to plant nomenclature, morphology, classification, ecology, geography, and history. The text is authoritative, and the book has been summed up as "unquestionably the outstanding book on plants of the year—in fact, of several years." [R: LJ, Nov. 1, 1970, p. 3793]

479. Everett, Thomas H. **Living Trees of the World**. New York: Doubleday, 1968. 351p.

Another convenient source for tree identification on a level with Brockman (entry 466).

480. Fink, Bruce. **The Lichen Flora of the United States**. Ann Arbor, Mich.: University of Michigan Press, 1935. 426p., plus 47 black-and-white plates.

This is out-of-print, but if it is available, it provides authoritative descriptions of 1,578 species, varieties, and forms belonging to 178 genera and 46 families of lichens. It is a basic work with an exhaustive treatment that laid the foundation for the study of lichens in the United States.

481. Fitter, Alastair, and Richard Fitter. **Collins Guide to the Grasses, Sedges, Rushes and Ferns of Britain and Northern Europe**. London: Collins; distr., Brattleboro, Vt.: Greene, 1984. 256p. ill. index. $5.50pa. ISBN 0002191369.

This source includes all native grasses, sedges, rushes, ferns, horsetails, and other fern allies of the northwestern quadrant of Europe. With *Wild Flowers of Britain and Northern Europe* by Richard Fitter, A. H. Fitter, and M. Blamey (4th ed., 1974), this book comprises a complete field guide to the higher plants of Northern Europe. Keys arranged by family, many illustrations, convenient size, distribution maps, and scientific and common name indexes make these books a very attractive, useful set.

482. **Flora Europaea**. New York: Cambridge University Press, 1964-80. 5v. $485.00/set. ISBN 0521232058.

482a. **Consolidated Index to Flora Europaea**. Compiled by G. Halliday and M. Beadle. New York: Cambridge University Press, 1983. 210p. $84.50. ISBN 0521224934.

Flora Europaea presents a synthesis of all the national and regional floras of Europe, based on critical reviews of existing literature and on studies in herbaria and in the field. It is an authoritative, unique resource. The *Consolidated Index* covers families, genera, and species for all five volumes, serves as a key to the numerical arrangement of the *Flora*, and documents the sources of all chromosome numbers.

While *Flora Europaea* is the indispensable reference in matters of nomenclature and taxonomy, distribution patterns are mapped by *Atlas Florae Europaeae: Distribution of Vascular Plants in Europe*, edited by Jaakko Julas and Juba Suominen, on the basis of studies by a team of European botanists. The *Atlas*, now extending to six volumes (Vol. 6, 1983), is published in Helsinki for the Committee for Mapping the Flora of Europe and the Societas Biologica Fennica Vanamo. Volume 1 was published in 1972. For information on sources of the chromosome numbers cited in *Flora Europaea* and for a basic checklist of recognized taxa, consult *Flora Europaea Check-List and Chromosome Index* by D. M. Moore (New York: Cambridge University Press, 1982, 423p., ISBN 0521237599).

483. **Flora Neotropica**. Monograph No. 1- . New York: New York Botanical Garden, 1964- . irregular. price varies. ISSN 0071-5794.

This important series of taxonomic accounts of plant groups or families growing spontaneously in the Americas between the Tropics of Cancer and Capricorn is the official publication of the Organization for Flora Neotropica. Data include ecology, cytology, anatomy, morphology, chemistry, economic importance, bibliography, and citation of specimens. The pagination for each monograph varies from a few pages to several hundred. The series is updated in *Taxon* (entry 193).

484. **Flora North America (FNA): Type Specimen Register**. Washington, D.C.: Smithsonian Institution, n.d.

This centralized databank of taxonomic information for North American flora is far from complete, although parts of it are available for the use of researchers. Information and reports of the project may be found in "The Flora North America Reports – A Bibliography and Index," *Brittonia* 29, no. 4 (1977): 419-32. See *North American Flora* (entry 532) for more information from a project with similar goals.

485. **Flora of Canada**. By H. J. Scoggan. Ottawa: National Museum of Natural Sciences; distr., Chicago: University of Chicago Press, 1978-79. 4v. $165.00/set. ISBN 066000139X. (National Museum of Natural Sciences Publications in Botany, No. 7).

Part 1 is a general survey with a glossary and references. Part 2 covers Pteridophyta, Gymnospermae, and Monocotyledoneae, including keys and a systematic section. Part 3 covers Dicotyledoneae (Saururaceae to Violaceae) with keys. Part 4 covers Dicotyledoneae (Loasaceae to Compositae) with keys. There is a cumulative index to Latin names of families, genera, and species for all four volumes.

486. **Flora of the Great Plains**. Edited by T. H. Barkley. The Great Plains Flora Association. Lawrence, Kans.: University Press of Kansas, 1986. 1408p. $55.00. ISBN 070060295X.

Flora of the Great Plains is the definitive reference work on plants of the Great Plains. A manual of the vascular plants that occur spontaneously in the region, it is the only up-to-date floristic treatment for the area. The *Flora* includes general keys, descriptions, range information, flowering times, ecological data, and synonymies for every flowering plant, conifer, and fern known to occur in the region.

487. **Flora SSSR**. Edited by V. L. Komarov. Moscow: Akademiia Nauk SSSR, 1934-60. 30v., plus index. (Vol. 11: repr., Forestburgh, N.Y.: Lubrecht and Cramer, 1985. $103.50. ISBN 3574292312).

This is a major regional flora, included here because of its importance and coverage. There is an English translation, *Flora of the USSR*, by the Israel Program for Scientific Translation, under the sponsorship of the National Science Foundation and the Smithsonian Institution.

488. Francki, R. I., Robert G. Milne, and T. Hatta. **Atlas of Plant Viruses**. Boca Raton, Fla.: CRC Press, 1985. 2v. bibliog. index. Vol. 1: $80.50. ISBN 0849365015. Vol. 2: $95.50. ISBN 0849365023.

This is a comprehensive collection of plant virus electron micrographs; there are 192 examples of all taxonomic groups designated by the International Committee on Taxonomy of Viruses. Data include information on physical, chemical, structural, cytopathological, and antigenic properties of viruses. This should be of great value for botanical viral research.

489. Frohne, Dietrich, and Hans Jurgen Pfander. **A Colour Atlas of Poisonous Plants: A Handbook for Pharmacists, Doctors, Toxicologists, and Biologists**. London: Wolfe; distr., Dobbs Ferry, N.Y.: Sheridan, 1984. 292p. $65.00. ISBN 0723408394.

Translated from the second German edition, this book aims to be comprehensive in scope for diagnosis and evaluation of poisoning by plants. It deals with indigenous and decorative plants of central Europe. Plants occurring in Britain have been given names recommended by the Botanical Society of the British Isles. An extensive bibliography and an index are provided. There are general discussions on problems from plant poisoning, toxicological significance, plant constituents, and an atlas of the most important plants with alleged or actual toxic properties. Color illustrations, identification keys, toxic constituents, symptoms of poisoning, treatment, and some examples from each family are included. Appendixes include tabular synopsis of berry-like fruits and compilation of leaf characters for the plants described. This professional reference manual is recommended for all botanical reference libraries.

490. Gabriel, Ingrid. **Herb Identifier and Handbook**. Adapted by E. W. Egan. New York: Sterling, 1975. 256p.

This guide to over 100 medicinal and culinary herbs provides descriptions, illustrations, cultivation techniques, harvesting, and preparation information. *ARBA* calls it a "most compact, informative and attractive reference guide." [R: ARBA, 1977, entry 1348]

491. Genders, Roy. **Scented Flowers of the World**. London: Robert Hale, 1977. 560p.

This fascinating book includes a wealth of hard-to-retrieve information of interest to the botanist as well as the hobbyist. It discusses the history, classification, and extraction of scents and presents an alphabetical guide to plants, trees, and shrubs with scented bark, leaves, flowers, and fruits.

492. Gleason, Henry A. **The New Britton and Brown Illustrated Flora of the Northeastern United States and Adjacent Canada**. rev. ed. New York: Hafner, 1975. 3v. ill. $115.00. ISBN 002845300X.

This monumental classic is comprehensive and reliable for all known species of seed plants, ferns, and fern allies.

493. Godfrey, Robert K., and Jean W. Wooten. **Aquatic and Wetland Plants of Southeastern United States: Dicotyledons**. Athens, Ga.: University of Georgia Press, 1981. 944p. $45.00. ISBN 0820305324.

Companion volume to *Aquatic and Wetland Plants of Southeastern United States: Monocotyledons* by the same authors. The aim of these volumes is to aid in identifying native and naturalized plants inhabiting aquatic and wetland places that are permanently or seasonally wet. This is an authoritative set that provides keys, description of species, habitat, and illustrations.

494. Gould, Frank W., and Robert B. Shaw. **Grass Systematics**. 2nd ed. College Station, Tex.: Texas A & M University Press, 1983. 397p. ill. bibliog. index. $25.00; $15.00pa. ISBN 089096145X; 0890961530pa.

Although written as a text in agrostology, the documentation and keys make this book appropriate for systematic grass research. Emphasis is placed on grass structure and growth and on the characteristics of U.S. grass genera. Appendixes discuss the preparation and handling of grass specimens and nomenclature of grasses. There is a glossary. For more information, see Hitchcock (entry 503).

495. **Gray's Manual of Botany**.
"A handbook of the flowering plants and ferns of the central and northeastern United States and adjacent Canada," this manual has long been the standard descriptive manual for flowering plants and ferns. For bibliographic information, see entry 556.

496. Grieve, Maud. **A Modern Herbal: The Medicinal, Culinary, Cosmetic and Economic Properties, Cultivation and Folk-Lore of Herbs, Grasses, Fungi, Shrubs and Trees, with All Their Modern Scientific Uses, with a New Service Index**. Magnolia, Mass.: Peter Smith, 1971. 2v. ill. $30.00/set. ISBN 0844603023.
This encyclopedic treatment gives synonyms, common and scientific names, part used, habitat, descriptions, medicinal action and uses, constituents, doses, poisons, history, and recipes as applicable. It is one of the most complete, modern herbals available, reporting on 800 varieties of plants. This and Harris (entry 501) are both well done, complementary to each other, and interesting.

497. Grout, Abel Joel. **Mosses with Hand Lens and Microscope**. 3rd ed. New York: Published by the author, 1924; repr., New York: Johnson, 1972. 339p. $15.00. ISBN 0910914036.
"Popular guide to the common or conspicuous mosses and liverworts of the North-Eastern United States." This can serve as an alternative to Conard in the *How to Know* (entry 505) series. Consult Smith (entry 546) for information concerning British mosses.

498. Hall, A. **The Wild Food Trailguide**. New York: Holt, 1976. 230p.
This book consists of drawings and descriptions of edible and poisonous wild plants and their collection and use. It is an appropriate and excellent choice for the layperson and most libraries. Hall can be used as an alternative to Angier's (entry 456) and Brouk's (entry 467) books on edible wild plants.

499. Hardin, James W., and James M. Arena. **Human Poisoning from Native and Cultivated Plants**. 2nd ed. Durham, N.C.: Duke University Press, 1973. 192p. $14.75. ISBN 0822303035.
This book is written for the layperson on a nontechnical level, presenting lists of plants that cause allergy, dermatitis, or internal poisoning. Included are a glossary, bibliography, index to scientific and common names, and a short chapter on poisoning of pets. It does not attempt to compete with Kingsbury (entry 507) or Lampe (entry 511) in treatment or scope; "the book is an accurate, handy reference for the layman." [R: AAAS, May 1975, p. 21]

500. Harrington, Harold Davis. **How to Identify Grasses and Grasslike Plants (Sedges and Rushes)**. Chicago: Swallow Press, 1977. 142p. $8.95pa. ISBN 0804007462.
Written by a professional botanist, this clear, concise, well-organized manual gives practical information for identifying grasses. There are 500 drawings and illustrations, with a section listing manuals and floras appropriate to the subject.

501. Harris, Ben Charles. **The Compleat Herbal: Being a Description of the Origins, the Lore, the Characteristics, the Types, and the Prescribed Uses of All Common Medicinal Plants**. Barre, Mass.: Barre Publishers, 1972. 243p. ISBN 0827172117; 0827172001pa.

This is an authoritative publication by a registered pharmacist who was also a Curator of Economic Botany of the Worcester Museum of Natural History. There is historical and philosophical information about herbal medicine, healing herbs, and herbal remedies. Harris is not as comprehensive as Grieve (entry 496), but some of the information may be more contemporary—the two may be used together. [R: ARBA, 1973, entry 1587]

502. Hillier, Harold G. **Hillier's Manual of Trees and Shrubs**. 5th ed. New York: Van Nostrand Reinhold, 1983. 600p. ill. $19.95. ISBN 0442236638.

This definitive reference for cultivated trees and shrubs provides descriptions for over 8,000 plants with color illustrations for more than 60 of the plants described. This British work is complementary to Elias (entry 475), Little (entry 514), and Sargent (entry 541).

503. Hitchcock, Albert Spear. **Manual of the Grasses of the United States**. 2nd ed. Washington, D.C.: U.S. Department of Agriculture, 1950; repr., Magnolia, Mass.: Peter Smith, 1971. 2v. $30.00/set. ISBN 0844603090. (U.S. Department of Agriculture Miscellaneous Publication No. 200).

This is the classic manual and definitive encyclopedia of all grasses known to grow in the continental United States, excluding Alaska. There are descriptions of 169 genera and 1,398 species, nearly all of them with illustrations. Maps, distribution, and keys to the tribes are included. For more recent treatment, see Gould (entry 494). [R: Phyto, Dec. 1972, p. 502]

504. Holm, Leroy, et al. **A Geographical Atlas of World Weeds**. New York: John Wiley, 1979. 394p. $53.50. ISBN 0471043931.

This record of world weed species in 124 countries is comprehensive for weeds and their distribution. Scientific names with up to three common synonyms are provided. The bulk of the book is a list of weeds, countries of distribution, and an indication of importance.

505. **How to Know Series**.

All of the volumes in this very successful series present introductory discussions on how to look, where to look, how to collect, and how to use keys. This series is written for the amateur botanist and includes glossaries, indexes, and numerous illustrations. Following are several good examples of the series:

Conard, Henry Shoemaker, and Paul L. Redfearn, Jr. **How to Know the Mosses and Liverworts**. 2nd ed. Dubuque, Iowa: William C. Brown Company, 1979. 320p. index. ISBN 0697047687.

Cronquist, Arthur. **How to Know the Seed Plants**. Dubuque, Iowa: William C. Brown Company, 1979. 250p. index. ISBN 069704760X.

Pohl, Richard Walter. **How to Know the Grasses**. 3rd ed. Dubuque, Iowa: William C. Brown Company, 1978. 208p. bibliog. index. ISBN 0697048764.

Prescott, Gerald Webber. **How to Know the Aquatic Plants**. 2nd ed. Dubuque, Iowa: William C. Brown Company, 1980. 180p. bibliog. indexes. ISBN 069704775X.

Wilkinson, Robert E., and Harry Edwin Jaques. **How to Know the Weeds**. 3rd ed. Dubuque, Iowa: William C. Brown Company, 1979. 235p. index. ISBN 0697047652. (spiral-bound).

506.   Hutchinson, John. **Key to the Families of Flowering Plants of the World**. rev. and enl. ed. Monticello, N.Y.: Lubrecht and Cramer, 1979. 117p. $35.00. ISBN 3874291618.

Intended as a supplement to *Genera of Flowering Plants*, this is one of the best-known keys to the flowering plants. Illustrations and a glossary are added to assist the botanical novice, although this key and Thonner's (entry 549) are more appropriate for the serious student than the amateur.

507.   Kingsbury, John M. **Poisonous Plants of the United States and Canada**. 3rd ed. Englewood Cliffs, N.J.: Prentice-Hall, 1964. 626p. $38.95. ISBN 0136850162.

Long the standard, this book updates and augments *Poisonous Plants of the United States* by W. C. Muenscher (1939). It was written for the physician or veterinarian and, in the final analysis, will probably suffer in comparison with the much newer book on poisonous plants by Lampe (entry 511) for the American Medical Association. Though authoritative for the time it was written, updated scientific information and a more convenient and attractive format would help it compete with newer manuals, like the *AMA Handbook* and Frohne (entry 489).

508.   Knobel, Edward. **Field Guide to the Grasses, Sedges and Rushes of the United States**. Revised by M. E. Faust. New York: Dover, 1977. 83p. ill. $2.75pa. ISBN 048623505X.

The small guide outlines a key to over 370 of the most common species of grasses, sedges, and rushes that is easy to carry, an excellent companion in the field. There are 500 line drawings with concise, accurate descriptions. [R: STBN, Nov. 1979, p. 1349]

509.   Krochmal, Arnold, and Connie Krochmal. **A Field Guide to Medicinal Plants**. New York: New York Times Book Company, 1984. 274p. ill. bibliog. indexes. $7.95pa. ISBN 0812963369.

Illustrated with more than 375 pictures for identification purposes, this book discusses 272 plants, giving common and scientific names, a brief description, habitat, season, harvesting, and uses. It carries a warning from its authors concerning the risky business of self-medication with wild plants. Taken in this context, the information is interesting, useful, and attractively presented. [R: ARBA, 1975, entry 1466]

510.   Kunkel, Gunther. **Plants for Human Consumption**. Königstein, Germany: Koeltz Scientific Books; distr., Forestburgh, N.Y.: Lubrecht and Cramer, 1984. 393p. $36.75. ISBN 3874292169.

"An annotated checklist of the edible phanerogams and ferns."

511.   Lampe, Kenneth F., and Mary Ann McCann. **AMA Handbook of Poisonous and Injurious Plants**. Chicago: American Medical Association; distr., Chicago: Chicago Review Press, 1985. 434p. ill. index. $18.95pa. ISBN 0899701833.

This authoritative reference is advertised as a convenient field guide to the identification, diagnosis, and management of human intoxications from plants and mushrooms of the United States, Canada, and the Caribbean. Designed for health care professionals, it is appropriate for all reference libraries. There are 437 full-color photographs to aid in identification, introductory materials pertaining to epidemiology of plant poisoning, botanical nomenclature, scientific and common name indexes, and sections on systemic plant poisoning, plant dermatitis, and mushroom poisoning. Each plant is discussed according to description, distribution, toxic part, toxin, symptoms, management, and references to the literature. This significant publication sponsored by the American Medical Association is the manual of choice when dealing with poisonous and injurious plants.

512. Lellinger, David B. **A Field Manual of the Ferns and Fern-Allies of the United States and Canada**. Washington, D.C.: Smithsonian Institution Press, 1985. 320p. ill. bibliog. index. $45.00; $29.95pa. ISBN 0874746027; 0874746035pa.

Written for the professional, amateur botanist, and horticulturist, this lavishly illustrated guide by an associate curator of botany at the Smithsonian Institution presents descriptions for 406 native and naturalized species, subspecies, and important varieties of ferns and fern-allies. Photographs, supplied by A. M. Evans, a professor of botany, include 402 color and 26 black-and-white illustrations. The book is indexed and includes a glossary and an extensive bibliography. [R: LJ, Feb. 15, 1986, p. 173-74]

513. Le Strange, Richard. **A History of Herbal Plants**. New York: Arco Publishing, 1977. 304p. $15.00. ISBN 0668042478.

This reference includes the botany, history, folklore, cultivation, glossary, bibliography and indexes to botanical and vernacular names for herbal plants. It is appropriate for a wide range of people who need a convenient, ready reference. The present volume is a "worthy successor" to Grieve (entry 496). [R: Choice, June 1978, p. 568]

514. Little, Elbert L., Jr. **Atlas of United States Trees**. Washington, D.C.: Government Printing Office, 1971-78. 5v., plus supplement. (Vol. 5: $4.25pa.).

This authoritative set provides tree distribution maps for species native to the continental United States. Dr. Little is chief dendrologist of the U.S. Forest Service, a specialist in the study of trees, and an authority on endangered species. This reference set, basic for any botanical or large public library, can be used with Little's *Checklist* (entry 559) to provide a balanced array of information about trees in the United States. [R: ARBA, 1978, entry 1305; Phyto, Sept. 1971, p. 128]

515. Lloyd, Francis E. **The Carnivorous Plants**. Waltham, Mass.: Chronica Botanica, 1942; repr., New York: Dover, 1976. 384p. ill. $7.95pa. ISBN 0486233219.

This historical review and definitive summary of the carnivorous or insectivorous plants is a thorough, scholarly treatment of 450 species, including fungi. A classic in its field, it is an essential reference. It may be used to complement Schnell (entry 542).

516. Loewenfeld, Claire, and Philippa Back. **The Complete Book of Herbs and Spices**. New York: Putnam, 1974. 313p. bibliog. index. ISBN 0399133738.

There are sections for general information and history, use, descriptions, and habitat for 138 species, cultivation, harvesting, storage, and recipes for herbs and spices of the world. The book gives comprehensive coverage from well-known authors. [R: ARBA, 1975, entry 1484]

517. Martin, William Keble. **The New Concise British Flora**. Nomenclature edited and revised by D. H. Kent. London: Michael Joseph and Ebury Press; distr., Lawrence, Mass.: Merrimack Book Service, 1982. 256p. ill. $25.00. ISBN 0718121260.

This superb book has 1,400 beautiful illustrations of British wildflowers with descriptions appropriate for identification purposes. The work is authoritative and accurate, and especially suitable for the person preferring pictures to keys. [R: ARBA, 1984, entry 1308]

518. McGinnis, Michael R., and Richard F. D'Amoto. **Pictorial Handbook of Medically Important Fungi and Aerobic Actinomycetes**. New York: Praeger Publishers, 1981. 172p. $22.95. ISBN 0030583640.

This handbook is a simple, complete guide for the identification of commonly encountered clinical laboratory molds, aerobic actinomycetes and yeasts. It is well illustrated with drawings and micrographs; the quality is high and the descriptions are concise with accurate pictorial references.

519. McIlvaine, Charles, and Robert K. Macadam. **One Thousand American Fungi: Toadstools, Mushrooms, Fungi, Edible and Poisonous with Full Botanic Descriptions**. rev. ed. New York: Dover, 1973. 729p. $9.95. ISBN 0486227820.

This reprint of the 1902 edition is the largest, most complete collection on American mushrooms written for the nonspecialist. There is information on size, shape, color, structure, texture, growing habits, size of spore, shape of gills, habitat, range, and season. A glossary, line drawings, edibility, antidotes for poisonous mushrooms, meanings of Latin names, and the historical background for many species are included in this old, but reliable guide. There is a new section on nomenclature by R. L. Shaffer, a recognized authority on mycology. Information on many uncommon species that is not available in many of the new guides is included in this source. [R: ARBA, 1974, entry 1521]

520. **Medicinal Plants of the World**. No. 1- . Algonac, Mich.: Reference Publications, 1978- . irregular.

The aim of this series is to collect widely dispersed data from different parts of the world into convenient sourcebooks for each region. For the interested general public and the scientific community, this series is also especially useful for phytochemists and pharmacologists. Although descriptions vary from book to book, data usually include selected list of floras of the region, scientific and common names, distribution, uses, and illustrations. The latest volume in the series is No. 4, *Medicinal Plants of China* by Edward S. Ayensu and James A. Duke, 1985 (2v., $94.95/set, ISBN 0917256204). [R: ARBA, 1986, entry 1487]

521. **Medicines from the Earth: A Guide to Healing Plants**. Edited by William A. R. Thomson. New York: McGraw-Hill, 1978. 208p. ISBN 0070560870.

Descriptions of 247 flowers and herbs useful for medicinal purposes are given. The book is abundantly illustrated and recommended for those seeking information on the art of herbal healing. [R: ARBA, 1979, entry 1354]

522. Medsger, Oliver P. **Edible Wild Plants**. New York: Macmillan, c1939, 1972. 359p. ill. $6.95pa. ISBN 0020809107.

Arranged by type of food, this volume discusses 150 important species of edible wild plants. The text is interesting and provides the necessary scientific information.

Medsger has stood the test of time well and can be compared favorably with Angier (entry 456), Hall (entry 498), and Brouk (entry 467). [R: AMN, Nov. 1939, p. 760]

523. Menninger, Edwin Arnold. **Flowering Vines of the World: An Encyclopedia of Climbing Plants**. New York: Hearthside Press, 1970. 410p.

Although the *ARBA* review of this book had some reservations about its quality, it is one of the few available books on flowering vines. It aims to be comprehensive and it is useful in many respects. For additional information, see Newcomb (entry 531). [R: ARBA, 1971, entry 1606]

524. Miller, Orson K., Jr. **Mushrooms of North America**. rev. ed. New York: Dutton, 1979. 359p. $16.95; $11.50pa. ISBN 052516166X; 052547482Xpa.

This is an excellent guide that, with Smith (entry 545), is the standard in the field. It is comprehensive, very well illustrated with 422 color photographs, and includes keys, glossary, bibliography, and index. The general introductory material is interesting and the guide is written for the amateur as well as the student of botany. This book is more appropriate for comparing specimens in the laboratory or at home, as its size is not conducive to field work. Miller and Smith are complementary resources but McIlvaine (entry 519) is more comprehensive. For a guide to poisonous mushrooms, see entry 454.

525. Millspaugh, Charles F. **American Medicinal Plants: An Illustrated and Descriptive Guide to Plants Indigenous To and Naturalized in the United States Which Are Used in Medicine**. New York: Boericke, 1887; repr., New York: Dover, 1974. 827p. $12.95. ISBN 0486230341.

This herbal, originally published in 1882, was written by a renowned physician and botanist. The classification and nomenclature have been revised, but many of the remedies are now in question. The book has historical interest; the contents may be checked with Harris (entry 501), Grieve (entry 496), and Duke (entry 474) for accuracy.

526. Moldenke, Harold Norman, and Alma L. Moldenke. **Plants of the Bible**. Waltham, Mass.: Chronica Botanica, 1952. 328p.

This book provides historical sketches, geographical descriptions, and scriptural citations for 230 plants mentioned in the Bible. Several plates are included as is a lengthy bibliography to provide authority for the work. This is the most scholarly and comprehensive treatment of the four included in this section; Anderson (entry 455), Walker (entry 551), and Zohary (entry 555) have other qualities that make them an attractive complement to Moldenke.

527. Morton, Julia Frances. **Atlas of Medicinal Plants of Middle America, Bahamas to Yucatan**. Springfield, Ill.: Charles C. Thomas, 1981. 1420p. bibliog. indexes. $147.50. ISBN 0398040362.

The *ARBA* review for this atlas calls it "unique and authoritative." It is indeed that. Written by a recognized scholar, it pulls together hard-to-find descriptions and illustrations for over 1,000 medicinal plants used in South America, the West Indies, the Bahamas, and Central America as far north as Yucatan. The audience for this resource will include scientists in anthropology, toxicology, pharmacology, and phytochemistry. [R: ARBA, 1983, entry 1328]

528. Morton, Julia Frances. **Folk Remedies of the Low Country**. Miami, Fla.: E. A. Seeman, 1974. 176p. $12.95. ISBN 0912458461.

Another book by the prolific J. F. Morton discusses plant remedies used in South Carolina. The work is well done with the usual dependable scientific descriptions and includes a color photograph of each plant. For students of pharmacognosy, this book has appeal and interest well beyond a strictly regional basis.

529. Morton, Julia Frances. **Major Medicinal Plants: Botany, Culture and Uses**. Springfield, Ill.: Charles C. Thomas, 1977. 431p. ill. bibliog. index. $62.75. ISBN 039803673X.

This authoritative book focuses on plants used for therapeutic purposes with emphasis on their botany, culture, and harvesting. Data are provided for nomenclature, taxonomy, description, origin and distribution, chemical constituents, propagation, culture, harvesting, medicinal uses, toxicity, and other uses. The information is dependable with illustrations provided for each entry. This excellent book from a respected researcher is appropriate for public and botanical libraries.

530. Muhlberg, Helmut. **The Complete Guide to Water Plants: A Reference Book**. rev. ed. East Ardsley, England: EP Publishing; distr., New York: Sterling, 1982. 392p. $14.95. ISBN 0715807897.

This well-produced book discusses aquatic plants in the wild: their cultivation, anatomy, physiology, propagation, nomenclature, and classification and provides short descriptions of 200 species. The book is appropriate for the serious student as well as the amateur hobbyist. It may be used with Prescott from the How to Know series (entry 505) and Godfrey (entry 493), if more information is required. Also, see *Water Plants of the World* (entry 552). [R: ARBA, 1983, entry 1316]

531. Newcomb, Lawrence. **Newcomb's Wildflower Guide: An Ingenious New Key System for Quick, Positive Field Identification of the Wildflowers, Flowering Shrubs and Vines of Northeastern and North Central North America**. Boston: Little, Brown & Co., 1977. 490p. $18.45. ISBN 0316604410.

Written for those with no formal botanical training, this book discusses 1,375 wildflowers, shrubs, and vines. The Newcomb system is based on answering five basic questions for diagnostic identification and relies on easily seen features that are unique. The book is not comprehensive. It has the official endorsement of the Garden Club of America.

532. **North American Flora**. Series II- . New York: New York Botanical Garden, 1954- . irregular. ISSN 0078-1312. (Pt. 12: 1984. $10.75. ISBN 0893272604).

The purpose of this series is to provide keys and descriptions of all plants growing spontaneously in North America, Central America, and the West Indies (excluding those islands whose flora is essentially South American). Series I was published from 1909 through 1949, issued at irregular intervals in parts numbered sequentially. Each part is devoted to an order, family or smaller group, and is complete with bibliography and index.

533. Perry, Frances, and Roy Hay. **A Field Guide to Tropical and Subtropical Plants**. New York: Van Nostrand Reinhold, 1982. 136p. $10.95; $6.95pa. ISBN 0442268610; 0442268599pa.

This "delightful" guide to more than 200 tropical plants is perfect to take on vacations. The descriptions are accompanied by color illustrations to aid in identification.

It is extremely useful and recommended for botanical collections and public libraries with a traveling clientele. [R: LJ, Mar. 1, 1983, p. 448]

534. **Peterson's Field Guide Series** and/or **Nature Library**.
This is a well-known and respected field guide series. The books are usually pocket-size, well-illustrated, and appropriate for students and amateurs. Following are a few examples of interest to botanists:

Brown, Lauren. **Grasses: An Identification Guide**. Boston: Houghton Mifflin, 1979. 240p. $9.95. ISBN 0395276241.

Peterson, Roger T., and Margaret McKenny. **A Field Guide to Wildflowers of Northeastern and North Central North America**. Boston: Houghton Mifflin, 1977. 420p. $14.95. ISBN 039508086X.

Petrides, George A. **A Field Guide to Trees and Shrubs**. 2nd ed. Boston: Houghton Mifflin, 1973. 428p. $15.95; $10.95pa. ISBN 0395136512; 0395175798pa.

535. Polunin, Oleg. **Trees and Bushes of Europe**. New York: Oxford University Press, 1976. 208p. ill. $18.95. ISBN 0192176315.
This author and this publisher, both well-known, have cooperated to produce an easy-to-use guide to the trees and bushes of Europe, with excellent photographs, drawings, and descriptions. There is a glossary, an appendix featuring photographs of tree barks, and a section on the uses of selected trees and bushes. See Hillier (entry 502) for complementary descriptions. [R: ARBA, 1978, entry 1307]

536. Polunin, Oleg, and Martin Walters. **A Guide to the Vegetation of Britain and Europe**. New York: Oxford University Press, 1985. 238p. ill. bibliog. index. $26.95. ISBN 0192177133.
This is billed as the first comprehensive account of the vegetation of Europe and Britain written for the amateur and general reader. It describes more than 100 plant communities and includes illustrations in color and black-and-white. The book includes distribution maps and a listing of national parks. [R: ARBA, 1986, entry 1489]

537. Preston, Richard J., Jr. **North American Trees (Exclusive of Mexico and Tropical United States): A Handbook Designed for Field Use, with Plates and Distribution Maps**. 3rd ed. Ames, Iowa: Iowa State University Press, 1976. 399p. $10.50. ISBN 0813811708.
This guide to trees is an alternative to several of the many tree books, such as Little (entry 514) and Sargent (entry 541). The beginner may need Petrides (see entry 534) as an identification companion. [R: ARBA, 1977, entry 1370]

538. Rickett, Harold W., and W. Niles. **Wild Flowers of the United States**. New York: New York Botanical Garden and McGraw-Hill, 1966-75. 6v., plus index. $325.00/set.
This outstanding set is a work of "beauty, scientific accuracy, and thoroughness." Written with the amateur botanist in mind, technical language is kept to a minimum. It is highly recommended for all botanical or public libraries. [R: Phyto, Mar. 1967, p. 390]

539.   Rinaldi, Augusto, and Vassili Tyndalo. **The Complete Book of Mushrooms**. Translated from the Italian by Italia and Alberto Mancinelli. New York: Crown Publishers, 1974. 330p. $14.98. ISBN 0517514931.

"Over 1,000 species and varieties of American, European, and Asiatic mushrooms with 460 illustrations in black-and-white and in color." This book for the amateur naturalist is a perfect complement for Miller (entry 524) and Smith (entry 545); its scope is broader than either of these to include common and easily identifiable species of foreign mushrooms.

540.   Sanecki, Kathleen Naylor. **The Complete Book of Herbs**. New York: Macmillan, 1974. 247p.

Yet another herb book that can be used as an alternative source for information on cultivation, use, herbs in cooking, history, and descriptive information. It is a practical book suitable for the gardener or the cook. There is more than the usual information on names in the appendix, which features the botanical name plus the common name of each herb in English, French, German, Spanish, and Italian.

541.   Sargent, Charles S. **Manual of the Trees of North America (Exclusive of Mexico)**. 2nd ed. Boston: Houghton Mifflin, 1922; repr., New York: Dover, 1965; repr., Magnolia, Mass.: Peter Smith, 1962. 2v. Vol. 1: $7.95pa. ISBN 0486202771; Vol. 2: $7.95pa. ISBN 048620278X. (Dover). $28.00/set. ISBN 0844628646. (Peter Smith).

This is the monumental work of a great dendrologist. It is comprehensive, authoritative, and the standard reference for native American trees. There are synoptic and analytical keys plus 100 other keys; 783 illustrations are included for 66 families with 185 genera and 717 species. Elias (entry 475) or Little (entry 514) may be considered an alternative; Brockman (entry 466) provides a handy field guide.

542.   Schnell, Donald E. **Carnivorous Plants of the United States and Canada**. Winston-Salem, N.C.: J. F. Blair, 1976. 125p. ill. (col.). $19.95. ISBN 0910244901.

This field guide to carnivorous green seed plants, appropriate for the beginning botanist, is intended for practical use. It does not have the scholarly treatment of Lloyd (entry 515) and is certainly not as complete, but it is effective on the amateur level. There are 117 full-color photographs.

543.   Silverman, Maida. **A City Herbal**. New York: Alfred A. Knopf, 1977. 181p. bibliog. index. ISBN 0394498526.

"A guide to the lore, legend and usefulness of 34 plants that grow in the city." A selected group of city plants, each with its own full-page line drawing, comprises this specialized herbal.

544.   **Simon and Schuster's Guide to Trees**. Edited by S. Schuler. New York: Simon and Schuster, 1977-78. 55p. $19.95; $9.95pa. ISBN 0671241249; 0671241257pa.

"A field guide to conifers, palms, broadleafs, fruits, flowering trees, and trees of economic importance," this volume also contains 300 descriptions of individual species on additional unnumbered pages. This guide is well illustrated with 650 illustrations, 350 of them in color. It may be used as an alternative or companion to Elias (entry 475), Sargent (entry 541), Preston (entry 537), Little (entry 514), or Hillier (entry 502).

545. Smith, Alexander H., and Nancy S. Weber. **The Mushroom Hunter's Field Guide**. Ann Arbor, Mich.: University of Michigan Press, 1980. 336p. $14.95. ISBN 0472856103.

This standard beginner's field guide to mushrooms is accurate and dependable, including simple keys, good illustrations, and authoritative descriptions. It is not as comprehensive as Miller (entry 524), but is, perhaps, easier to use. The *ARBA* reviewer suggested that a careful mushroom hunter should have access to both Smith and Miller. [R: ARBA, 1982, entry 1466]

546. Smith, Anthony John Edwin. **The Moss Flora of Britain and Ireland**. New York: Cambridge University Press, 1978. 706p. ill. $99.50. ISBN 0521216486.

This definitive manual can serve for identification, as a text, or for ready reference for the 692 species that are described and splendidly illustrated. It is appropriate for the amateur or professional botanist and is a major contribution to British botany. Refer to Conrad (entry 505) and Grout (entry 497) for identification of United States mosses. [R: QRB, June 1979, p. 185]

547. **Sturtevant's Edible Plants of the World**. Edited by U. P. Hedrick. Albany, N.Y.: New York Agricultural Experiment Station, 1919; repr., New York: Dover, 1972. 686p. $10.95pa. ISBN 0486204596.

Although dated, this unabridged reprint of the 1919 version is a valuable addition to the plant literature. The scope is worldwide and the bibliography is detailed. "There is a wealth of valuable material well organized here and therefore especially welcomed in this inexpensive reprinting." [R: Phyto, Sept. 1974, p. 72]

548. Symonds, George W. D. **The Shrub Identification Book**. New York: William Morrow, 1963. 379p. $17.95; $12.95pa. ISBN 0688000401; 0688050409pa.

Still in print, this identification guide obviously has proved its usefulness over the years. Its subtitle, "The visual method for the practical identification of shrubs, including woody vines and ground covers," accurately reflects its purpose. It is coded and arranged for easy use and includes pictorial keys, descriptions, bibliography, glossary, and an index. It could be used as a companion for Polunin (entry 535) and Hillier (entry 502).

549. **Thonner's Analytical Key to the Families of Flowering Plants**. Translation of **Anleitung zum Bestimmung der Familien der Blutenpflanzen (Phanerrogames)**. 2nd ed. By Franz Thonner. Berlin: n.p., 1917; repr., The Hague: Leiden University Press; distr., Hingham, Mass.: Kluwer Boston, 1981. 206p. $37.00; $21.00pa. ISBN 9060214617; 906021479Xpa.

This is one of the few keys to the families of flowering plants and can be substituted, in some respects, for Hutchinson (entry 506) and Newcomb (entry 531). It has recently been translated from the German and brought up-to-date so that it may be used with Willis's eighth edition (entry 310) and Hutchinson's third edition (entry 586).

550. Trelease, William. **Winter Botany**. 3rd ed. Urbana, Ill.: Published by the author, 1931; repr., New York: Dover, 1967. 393p. ill. $6.95pa. ISBN 0486218007.

This key to over 1,000 common trees and shrubs without foliage covers the northern United States and some southern species. All important information is provided in this classic work. It may be used in conjunction with Brown (entry 468).

551. Walker, Winifred. **All the Plants of the Bible**. London: Butterworth, 1957; repr., New York: Doubleday, 1979. 244p. $15.95. ISBN 0385149646.

There are watercolor illustrations for the more than 100 plants described in this attractive manual. This is a popular account written for the amateur with little scientific descriptive information. Anderson (entry 455) is more comprehensive, but the two are formatted in such different ways that they complement each other very usefully. For a more scholarly and comprehensive treatment, see Moldenke (entry 526) or Zohary (entry 555). [R: ARBA, 1981, entry 1423]

552. **Water Plants of the World: A Manual for the Identification of the Genera of Freshwater Macrophytes**. By Christopher D. K. Cook, et al. The Hague: Junk; distr., Hingham, Mass.: Kluwer Academic, 1974. 561p. ill. index. $125.00. ISBN 9061930243.

This manual helps to identify freshwater macrophytes on a worldwide scale, including all Charophyta, Bryophyta, Pteridophyta, and Spermatophyta. Two general identification keys are included; the work is arranged systematically.

553. **World Pollen Flora**. Edited by G. Erdtman. New York: Hafner, 1970-71. 4v. $39.95/set(pa.) ISBN 0028442105.

Taxonomy, pollen diagnosis, methods, measurements, and systematic relationships are discussed in this set describing pollen grains and spores.

554. **The Yeasts – A Taxonomic Study**. 3rd rev. and enl. ed. Edited by N. J. W. Kreger-van Rij. New York: Elsevier Scientific Publishing, 1984. 1082p. ill. bibliog. index. $181.25. ISBN 0444804218.

This revision, which provides the criteria and methods for classification and identification of yeasts, follows the format of the successful earlier editions. The first chapter discusses general classification; chapter 2 deals with methods for isolation, maintenance, classification, and identification; and the remaining chapters present keys for each genus, a standard description of each species including recognized varieties within a genus. A glossary, bibliography, and an index of taxa are provided. Although this volume is included in this section for identification purposes, this important treatise also represents a stage in the evolution of yeast classification and the study of yeast ecology, evolution, and speciation. This and Barnett (entry 461) are worthy competitors. [R: IJSB, Apr. 1985, p. 226]

555. Zohary, Michael. **Plants of the Bible: A Complete Handbook to All the Plants with 200 Full-Color Plates Taken in the Natural Habitat**. New York: Cambridge University Press, 1982. 224p. ISBN 0521249260.

Plants of the Bible are shown in their natural habitat; identifications and descriptions are authoritative and relevant citations from the Bible accompany the discussion of each plant. This book will be fascinating to nature lovers and Bible readers alike.

Checklists/Manuals of Names

These lists are useful for verifying botanical names. Following the listing of general manuals is a special listing of references on endangered species. It also may be necessary to consult several dictionaries, taxonomic indexes, floras, and manuals during the search for valid plant names.

General Checklists and Manuals

556. **Gray's Manual of Botany**. 8th ed. Rewritten and expanded by Merrit Lyndon Fernald. New York: American Book Co., 1950. 1632p.

"A handbook of the flowering plants and ferns of the central and northeastern United States and adjacent Canada." This is a standard, time-honored, reference source for the names and descriptions of plants. See also entry 495.

557. Hitchcock, Albert Spear. **Manual of the Grasses of the United States**.

This, like Gray (entry 556), is much more than a list of names, although it is useful in this regard. See entry 503 for bibliographic information with a more complete annotation.

558. Kartesz, John T., and Rosemarie Kartesz. **A Synonymized Checklist of the Vascular Flora of the United States, Canada, and Greenland**. Vol. II: **The Biota of North America**. Chapel Hill, N.C.: University of North Carolina Press, 1980. 494p. $35.00. ISBN 0807814229.

This work provides names and synonyms, in current use, of native or naturalized plants found in North America. Over 3,000 primary references were scanned to provide over 57,000 documented names with approximately 20,000 synonyms. One of the best sources for verifying names and locating synonyms, it is appropriate for specialized and large public libraries. [R: ARBA, 1981, entry 1431]

559. Little, Elbert L., Jr. **Checklist of United States Trees (Native and Naturalized)**. Washington, D.C.: Forest Service, U.S. Department of Agriculture, Government Printing Office, 1979. 375p. (Agriculture Handbook, No. 541).

This checklist provides accepted scientific names and current synonyms, approved common names, and geographic ranges of the native and naturalized trees of the United States, including Alaska, but not Hawaii. This list represents the official standard for tree names used by the Forest Service; it is recommended for academic and large public libraries. [R: ARBA, 1981, entry 1455]

560. Miller, Orson K., Jr., and David F. Farr. **An Index of the Common Fungi of North America (Synonymy and Common Names)**. Vaduz, Liechtenstein: J. Cramer; distr., Forestburgh, N.Y.: Lubrecht and Cramer, 1985. 206p. $15.00pa. ISBN 3768209741.

The purpose of this book is to list under one species name all the common and scientific names given over the years to a single species of fungi. Names are taken from 30 sources to produce 4,000 names, in more than 350 genera among 78 families of mushrooms. This list provides the means to discover if a species listed under different scientific and common names in different books is actually considered to be the same. This reference work is a unique and extremely useful source book.

561. **National List of Scientific Plant Names**. Washington, D.C.: U.S. Department of Agriculture, Soil Conservation Service, 1982. 2v. GD:A-57-15-159-v-1-2. SCS-TP-159.

This list contains symbols for scientific names; accepted names for genera, species, subspecies, and varieties; authors of plant names; symbols for source manuals and family names; symbols for plant habits; and symbols for regions of distribution. It includes the United States, Canada, and the Caribbean region. Volume 1 lists plant names; volume 2 provides synonymy and includes names and symbols that have been incorrectly used as well as directing users to the accepted name. This is a valuable addition to the collection of research and large public libraries.

562.   **A Provisional Checklist of Species for Flora North America**. rev. ed. Edited by Stanwyn G. Shetler and Laurence E. Skog. St. Louis, Mo.: Missouri Botanical Garden, 1978. 199p. (Monographs in Botany from the Missouri Botanical Garden, Vol. 1; Flora North America Report 84).

This list updates *Flora North America Report 64* to include taxon name, author name, plant characteristics, region, and source of species. It is a useful source of information that will be enlarged to form a comprehensive North American list of vascular plants.

563.   Scott, Thomas G., and Clinton H. Wasser. **Checklist of North American Plants for Wildlife Biologists**. Washington, D.C.: Wildlife Society, 1980. 58p. $5.00pa. ISBN 0933564074.

This official publication of the Wildlife Society was compiled to "facilitate communication among wildlife biologists." It lists scientific plant names published in three of the most important wildlife journals, arranged alphabetically, with information on common names where available, person who first described the plant, etc. This list will be useful for academic and botanical libraries. [R: ARBA, 1981, entry 1436]

564.   Terrell, Edward E. **A Checklist of Names for 3,000 Vascular Plants of Economic Importance**. Washington, D.C.: U.S. Department of Agriculture, 1977. 201p. $3.50pa. GD:A-1-76-505. (Agriculture Handbook, No. 505).

This reference provides two alphabetic checklists for scientific and common names of some 3,000 plants and is useful for verifying or identifying particular botanical names with their more popular versions. [R: ARBA, 1979, entry 1340]

Endangered Plants

565.   Ayensu, Edward S., and Robert A. DeFilipps. **Endangered and Threatened Plants of the United States**. Washington, D.C.: Published jointly by the Smithsonian Institution and the World Wildlife Fund, 1978. 403p. bibliog. $35.00. ISBN 0874742226.

This publication introduces the concepts of "endangered and threatened," discusses habitats and conservation, and provides lists of endangered and extinct plant species in the United States. Bibliographies, descriptions of methods, examples of computer programs, and maps are included.

566.   **Canada's Threatened Species and Habitats**. Edited by Theodore Mosquin and Cecil Suchal. Ottawa: Canadian Nature Federation, 1977. 185p.

This volume reports the proceedings of a symposium on Canada's threatened species and habitats.

567.   **Conservation of Threatened Plants**. Edited by J. B. Simmons, et al. New York: Plenum, 1976. 352p. $49.50. ISBN 0306328011.

Proceedings of the Conference on the Functions of Living Plant Collections in Conservation and Conservation-Orientated Research and Public Education, held at the

Royal Botanic Gardens, Kew, England, September 2-6, 1975, and sponsored by the NATO Special Program Panel on Eco-Sciences.

568. **Extinction Is Forever**. Edited by Ghillean T. Prance and Thomas S. Elias. New York: New York Botanical Garden, 1977. 437p. bibliog. index. $20.00. ISBN 0893271969.

The proceedings of a symposium held at the New York Botanical Garden, May 11-13, 1976, this volume covers "threatened and endangered species of plants in the Americas and their significance in ecosystems today and in the future." [R: Choice, Jan. 1978, p. 1523]

569. **IUCN Plant Red Data Book**. New York: Unipub, 1978. 540p. $20.00. ISBN 2880322022.

This publication includes information and lists of endangered plant species.

570. Kartesz, John T., and Rosemarie Kartesz. **Rare Plants**. Pittsburgh, Pa.: Biota of North America Committee, 1977. 361p. (Biota of North America, Pt. I: Vascular Plants).

Covers a variety of rare plants.

571. Koopowitz, Harold, and Hilary Kaye. **Plant Extinction: A Global Crisis**. Washington, D.C.: Stone Wall Press; distr., Harrisburg, Pa.: Stackpole Books, 1983. 256p. $16.95. ISBN 0913276448.

This book presents a popular discussion of the problems and ethics of plant extinction and includes case histories of selected plants. Appendixes provide lists of conservation organizations, Red Data lists of plants, references, periodicals, and there is an index.

572. **List of Rare, Threatened and Endemic Plants in Europe**. 2nd ed. Edited by Threatened Plants Unit (IUCN Conservation Monitoring Centre). Strasbourg, France: Council of Europe, 1983. 360p. £6.00pa. ISBN 9287102147. (Nature and Environment Series, No. 27).

This report lists threatened plants throughout Europe.

573. Miasek, Meryl A., and Charles R. Long. **Endangered Plant Species of the World**. . . . See entry 52.

574. Mohlenbrock, Robert H. **Where Have All the Wildflowers Gone? A Region-by-Region Guide to Threatened or Endangered U.S. Wildflowers**. New York: Macmillan, 1983. 239p. $15.95. ISBN 002585450X.

Arranged by region, this describes 60 wildflowers and 60 rare plants on the endangered list or under review for the endangered list.

575. **Report on Endangered and Threatened Plant Species of the United States**. Presented to the Congress of the United States of America by the Secretary, Smithsonian Institution. Washington, D.C.: Government Printing Office, 1975. 200p. (Serial No. 94-A).

This is an official report on endangered plants in the United States.

Classification, Nomenclature, and Systematics

This section includes sources for locating nomenclatural and botanical classification information. Both *Mycotaxon* (entry 169) and *Taxon* (entry 193) provide current update information on questions of systematics and nomenclature.

576. Becker, Kenneth M. "A Comparison of Angiosperm Classification Systems." **Taxon** 22, no. 1 (1973): 19-50.
Comparisons are made between the classification systems proposed by Cronquist, Takhtajan, Thorne, Engler, Hutchinson, and Bentham and Hooker. See Swift's *Botanical Classification* (entry 592) for additional comparisons.

577. Benson, Lyman. **Plant Classification**. 2nd ed. Lexington, Mass.: Heath, 1979. 901p. $31.95. ISBN 0669014893.
This elementary text is included here because it provides an introduction to the classification of living plants. It discusses vocabulary, plant characteristics for identification, preparation and preservation of specimens, the basis for classification, and the association of species in natural vegetation.

578. Cronquist, Arthur. **An Integrated System of Classification of Flowering Plants**. New York: Columbia University Press, 1981. 1152p. $132.00. ISBN 0231038801.
This is an indispensable work for botanical classification and an excellent compendium of information on the division, class, order, family, and basic features of 383 families in 83 orders of flowering plants. It can be used in place of Hutchinson (entry 586). Many critics believe that it will be the definitive work in the field for many years. [R: Choice, Jan. 1982, p. 647]

579. **Databases in Systematics**. Edited by R. Allkin and F. A. Bisby. New York: Academic, 1984. 329p. index. $49.50. ISBN 0120530406. (The Systematics Association Special Volume 26).
Contents include information on electronic data processing; taxonomic information services; European taxonomic, floristic, and biosystematic documentation systems; the IUCN database; and other international databases of interest to taxonomists. Although this information is dated, it should be useful in academic and professional libraries for review purposes.

580. Gibbs, R. Darnley. **Chemotaxonomy of Flowering Plants**. Montreal: McGill-Queen's University Press; distr., Buffalo, N.Y.: University of Toronto Press, 1974. 4v. bibliog. $150.00/set. ISBN 0773500987.
This highly technical book, which surveys plant chemistry as it pertains to plant taxonomy, contains succinct summaries of a huge amount of chemical literature. It is useful on several levels for the professional as well as the layperson. There are sections on the history of chemotaxonomy, criteria used in taxonomy, plant constituents, families and orders of dicots and monocots, a lengthy bibliography, and an "exhaustive compilation of all the chemistry of all the species that have been investigated. . . ." [R: ARBA, 1975, entry 1476]

581. Gledhill, D. **The Names of Plants**. New York: Cambridge University Press, 1985. 159p. bibliog. $34.50; $9.95pa. ISBN 0521305497; 052131562Xpa.

Part 1 of this discussion provides an account, for the amateur gardener of how and why the naming of plants has changed with time. Part 2, the glossary, is a translation and interpretation of the scientific names of plants from all over the world. Rules of botanical nomenclature and terminology and the significance of the international codes form an integral part of this reference; there is no information about people or places epitomized in plant names.

582. Harborne, Jeffrey B., and Billie Lee Turner. **Plant Chemosystematics**. Orlando, Fla.: Academic, 1984. 562p. bibliog. indexes. $95.00. ISBN 0123246407.

Although this is a textbook and not a reference source as such, it does provide a wealth of data and serves as a review of the state and potential of plant chemosystematics through 1982. There is an extensive bibliography; the literature relating to zoological and prokaryotic systematics is not included. There are indexes to subject and plant genera and species. [R: QRB, Dec. 1985, pp. 506-7]

583. Hawksworth, D. L. **Mycologist's Handbook**. Kew, England: Commonwealth Mycological Institute, c1974, 1981. 231p. $21.95. ISBN 0851983004.

"An introduction to the principles of taxonomy and nomenclature in the fungi and lichens." Useful for mycologists, this introduction to systematics includes extensive references, an index, abbreviations for serial publications, a glossary, and mycological examples illustrating the proper use of the International Code of Botanical Nomenclature.

584. Heywood, Vernon Hilton. **Plant Taxonomy**. 2nd ed. London: Edward Arnold, 1976. 64p. ISBN 713126086; 713126094pa. (The Institute of Biology's Studies in Biology, No. 5).

This volume provides clear, easily understood coverage of taxonomy for the student or nonspecialist. There are discussions of taxonomic structure, plant characters, biochemical systematics, computers in the study of taxonomy, plant evolution, and suggestions for further reading for the enthusiastic beginner.

585. Holmes, S. **Outline of Plant Classification**. New York: Longman, 1983. 192p. $17.95. ISBN 0582446481.

This clear and comprehensive guide to the classification of the plant kingdom is useful for the amateur botanist as well as the professional biological writer. The appendix provides flowcharts of plant life cycles for alternation of generations. It is authoritative, yet readable.

586. Hutchinson, John. **The Families of Flowering Plants Arranged According to a System Based on Their Probable Phylogeny**. 3rd ed. New York: Oxford University Press, 1973; repr., Monticello, N.Y.: Lubrecht and Cramer, 1979. 968p. $63.00. ISBN 387429160X.

This outstanding contribution to the study of botanical classification is a scholarly work appropriate for all botanical libraries. This discussion records influential systematic thinking that has been largely replaced by Cronquist's *Integrated System* (entry 578). [R: ARBA, 1975, entry 1477]

587.   **International Code of Botanical Nomenclature: Adopted by the Thirteenth International Botanical Congress, Sydney, Australia, 1981.** Edited by Edward G. Voss, et al. Utrecht, Netherlands: Bohn, Scheltema and Holkema, 1983. 488p. $49.25. ISBN 9031305723. (Regnum Vegetabile, Vol. 111).

Botanical nomenclature is governed by the International Code as adopted by each International Botanical Congress. The code aims at the provision of a stable method of naming taxonomic groups, avoiding and rejecting the use of names that may cause error, ambiguity, or confusion. Updates to the code may be found in *Taxon* (entry 193) or *Mycotaxon* (entry 169). This code also governs the use of scientific or Latin names for plants whether they are cultivated or wild. See the *International Code of Nomenclature for Cultivated Plants* (entry 588), which is revised after each new *International Code of Botanical Nomenclature* appears.

588.   **International Code of Nomenclature for Cultivated Plants, 1980.** Edited by C. D. Brickell. Utrecht, Netherlands: Bonn, Scheltema and Holkema, 1980. 32p. $8.75. ISBN 9031304468. (Regnum Vegetabile, Vol. 104).

This code governs nomenclature for cultivated plants. It was formulated and adopted by the International Commission for the Nomenclature of Cultivated Plants of the International Union of Biological Sciences.

589.   Jeffrey, Charles. **An Introduction to Plant Taxonomy.** 2nd ed. New York: Cambridge University Press, 1982. 100p. $24.95; $13.95pa. ISBN 0521245427; 0521287758pa.

This book discusses the fundamentals and process of classification, taxonomic hierarchy, naming of plants, systems of classification, and includes an outline of plant classification and an index. It is suitable as an understandable overview of taxonomy for beginning students and the general public.

590.   Sneath, Peter H. A., and Robert R. Sokal. **Numerical Taxonomy: The Principles and Practice of Numerical Classification.** San Francisco: W. H. Freeman & Co., 1973. 573p. $43.95. ISBN 0716706970.

This text discusses the aims and principles of numerical taxonomy.

591.   Stebbins, G. Ledyard. **Flowering Plants: Evolution above the Species Level.** Cambridge, Mass.: Harvard University Press, 1974. 480p. $30.00. ISBN 0674306856.

This is basically a synthesis of *The Evolution and Classification of Flowering Plants* by A. Cronquist (Boston: Houghton Mifflin, 1968) and Takhtajan (entry 593). This volume epitomized the consensus of prevailing thinking in 1974 on the classification of flowering plants.

592.   Swift, Lloyd H. **Botanical Classification: A Comparison of Eight Systems of Angiosperm Classification.** Hamden, Conn.: Archon, 1974. 374p.

Comparisons are made between the classification systems proposed by Endlicher (1836), Bentham and Hooker (1862), Eichlar (1874), Engler and Prantl (1900), Bessey (1915), Hutchinson (1959), Melchior (1964), and Cronquist (1968). See Becker (entry 576) for additional comparisons.

593.   Takhtajan, A. L. **Flowering Plants: Origin and Dispersal.** Translated by C. Jeffrey. Monticello, N.Y.: Lubrecht and Cramer, 1981. 310p. $28.00. ISBN 0050017152.

This reprint of the 1969 edition written by a Soviet authority on plant taxonomy, classification and origin of flowering plants is useful when considered in conjunction with Cronquist (entry 578).

594.   Thorne, R. "A Phylogenetic Classification of the Angiospermae." **Evolutionary Biology** 9 (1976): 35-106.
This synopsis represents Thorne's interpretations of the relationships of the major taxa at this point in geological time.

595.   **Vascular Plant Systematics**. Albert E. Radford, et al. New York: Harper & Row, 1974. 891p. ISBN 0060453087; 0060453095pa.
This book serves as a "how to" and "why do"; in other words, as a practical reference text and source book for the study of taxonomy. It discusses nomenclature; characters and evidence; floristics; data collection, analysis, presentation, and documentation; plant identification; systems of classification; and information on botanical literature, herbaria, botanical gardens, societies, and the like. It is updated by *Fundamentals of Plant Systematics* (New York: Harper & Row, 1986, 498p., $35.50, ISBN 0060453052).

596.   Wiley, Edward O. **Phylogenetics: The Theory and Practice of Phylogenetic Systematics**. New York: John Wiley, 1981. 439p. $45.50. ISBN 0471059757.
This has been described as a major work explaining the dominant position of Hennigian methods of phylogenetic systematics. It discusses the phylogenetic application of systematics research to the study of the pattern and process of evolution. [R: QRB, Mar. 1983, pp. 60-61]

597.   **The Yeasts—A Taxonomic Study**. See entry 554.

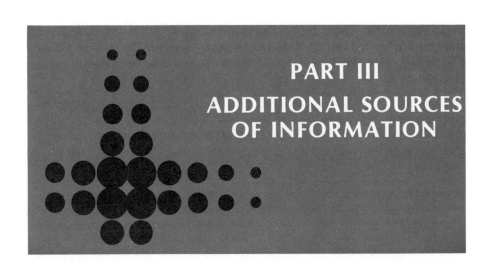

PART III
ADDITIONAL SOURCES
OF INFORMATION

8 Historical Materials

This section consists primarily of books introducing the history of the botanical sciences. There is no attempt to be comprehensive; focus is on the general, rather than on specific societies, subject areas, or people. Fossil history is not included. Most selections are in English and retrospective biographical materials concentrate on American and British sources. Examples include classic histories, textbooks, biographical dictionaries, bibliographies, as well as other sources that suggest additional reading; there is emphasis on early American reference materials.

The journals *Archives of Natural History, Bulletin of the Hunt Institute for Botanical Documentation, History and Philosophy of the Life Sciences, Huntia, ISIS,* and the *Journal of Natural History* all publish articles discussing the current thinking in the history of botany, although articles of a historical nature appear in a wide variety of journals. Entry to the historical scientific literature may be gained through *Biological Abstracts.* See chapter 2, "Abstracts, Indexes, and Databases," for more information on access to scientific and popular botanical articles appearing in journals or magazines.

On a regular schedule, the journal *ISIS* publishes its "Critical Bibliography of the History of Science," another satisfactory access point for botanical historians. *Taxon* publishes a very useful continuing series on well-known botanists, which includes their portraits and biographical data. In addition to sources listed here, consult chapter 1 for bibliographies relevant to the history of botany.

598.   Ainsworth, Geoffrey Clough. **Introduction to the History of Mycology**. New York: Cambridge University Press, 1976. 350p. ill. bibliog. indexes. $54.50. ISBN 0521210135.

This volume presents a documented outline of the development of the main branches of mycology, by tracing important themes from early times to the present. Chapters are mostly self-contained to present discussions on form and structure, classification, and pathogenicity, to name a few examples. There is a wealth of information in this scholarly, fascinating treatment. [R: QRB, Dec. 1977, p. 430]

599.   Ainsworth, Geoffrey Clough. **Introduction to the History of Plant Pathology**. New York: Cambridge University Press, 1981. 220p. bibliog. index. $75.00. ISBN 0521230322.

Although this topic is in the applied area and so, strictly speaking, should not be included in this guide to the literature, it was included here because of its more general interest to the historical study of botany as a whole. The treatment is authoritative and the comprehensive bibliographical and biographical references provide invaluable additional historical detail.

600.   **"And Some Brought Flowers": Plants in a New World**. Edited by Mary Alice Downie and Mary Hamilton. Buffalo, N.Y.: University of Toronto Press, 1980. 160p. bibliog. $24.95. ISBN 0802023630.

The reviewer called this an attractive book of unusual travel and exploration writing. There are 70 exquisite watercolor illustrations of plants, quotations from early explorers and settlers, and 20 pages of short biographies of men and women discussed in the text. [R: Huntia 5, no. 1, 1983, p. 77]

601.   Anderson, Frank J. **An Illustrated History of the Herbals**. New York: Columbia University Press, 1977. 269p. index. $32.00; $11.95pa. ISBN 0231040024; 0231083807pa.

This is a herbal sampler portraying the unique character and flavor of selected herbals by illustrations from the herbals themselves. For amateur botanists and college/public libraries. For a more comprehensive discussion, consult Arber (entry 602).

602.   Arber, Agnes. **Herbals: Their Origin and Evolution: A Chapter in the History of Botany 1470-1670**. 2nd ed. Cambridge, England: Cambridge University Press, 1953. 326p. ill. bibliog. index.

This standard history of herbals is an authoritative, scholarly discussion of the early history of botany, the earliest herbals, the botanical renaissance, and the evolution of the art of plant description, classification, and illustration. Appendixes include a chronological list of the principal herbals, a list of the historical and critical works consulted, and a subject index.

603.   Asmous, V. C. "Fontes Historiae Botanicae Rossicae." **Chronica Botanica** 11, no. 2 (1947): 87-118.

This article enumerates resources concerning the history of botany in Russia, including biographical and bibliographical publications.

604. **Bibliography and Natural History: Essays Presented at a Conference Convened in June, 1964**. Edited by T. R. Buckman. Lawrence, Kans.: University of Kansas Libraries, 1966. 148p.

Specialists from all over the world prepared papers for this conference on the bibliography of natural history. The chief emphasis was botanical, to include papers from such prominent historians and bibliographiers as W. T. Stearn, F. A. Stafleu, J. Stannard, S. Lindroth, and R. C. Rudolph, concentrating on the eighteenth and early nineteenth century in Europe and America. "Early American Botany and Its Sources" by J. Stannard (pp. 73-102) and "Two Centuries of Linnaean Studies" by S. Lindroth (pp. 27-45) are examples of two particularly useful studies containing valuable bibliographies for additional reading.

605. **Biographical Dictionary of Botanists Represented in the Hunt Institute Portrait Collection**. Hunt Botanical Library, Carnegie-Mellon University. Boston: G. K. Hall, 1972. 451p. $30.00. ISBN 0816110239.

The listings in this catalog include 11,000 people representing 17,000 portraits. Biographical information is added, as well as botanical speciality, and countries of principal activity. Other sources for portraits are Barnhart's *Biographical Notes* (entry 608), *Taxon* (entry 193), and Stafleu and Cowan's *Taxonomic Literature* (entry 117).

606. **Biographical Memoirs of Fellows of the Royal Society**. London: The Royal Society of London, 1955. 9v. (Vols. 1-9: 1932-54).

This reference includes portraits and biographical information for deceased members of the Royal Society of London, which unlike the U.S.'s National Academy, allows foreign membership. Consult volume 75 of the *Proceedings of the Royal Society* for obituary notices prior to 1932.

607. **Biographical Memoirs of the National Academy of Sciences**. Vol. 1- . Washington, D.C.: National Academy of Sciences, 1877/79- . annual. ISSN 0077-2933. (Vol. 54: 1983. $13.00).

This annual publication includes complete biographical information about the deceased members of the National Academy of Sciences. A portrait, bibliography of publications, and a complete chronology are provided for each scientist.

608. **Biographical Notes upon Botanists**. The New York Botanical Library. Compiled by J. H. Barnhart. Boston: G. K. Hall, 1965. 3v. $340.00/set. ISBN 0816106959.

This unique and invaluable source lists biographical details of botanists from the earliest times to the late 1940s, including information on their life, academic history, obituary notices, location of portraits, travels, and collections. See *Biographical Dictionary of Botanists* (entry 605) for similar information.

609. Blunt, Wilfrid J. W., with the assistance of William T. Stearn. **The Art of Botanical Illustration**. 4th ed. London: Collins, 1974. 304p.

Lavishly illustrated, this scholarly book deserves its reputation as the definitive and exhaustive historical survey of botanical illustration from prehistoric to modern times. It appeals to botanists, artists, amateur collectors, and students: a true gem. Appendixes include a series of eight articles on botanical drawing, some illustrated books on British plants, and sources of further information. [R: Phyto, Sept. 1975, pp. 501-2]

610.   Boivin, Bernard. "A Basic Bibliography of Botanical Biography and a Proposal for a More Elaborate Bibliography." **Taxon** 26, no. 1 (1977): 75-105.

This article is in two parts, as its title suggests. The first part discusses the design for an "analytical bibliography of collections of biographies, bibliographies and bio-bibliographies of botanical import." The second part is an extremely useful basic bibliography annotating over 200 sources of biographical information about botanists. Sources are grouped by Boivin under the headings "collectors," "herbaria," "biographies," and "regional collections." There is an index to this article in the November 1977 issue of *Taxon* (pp. 603-11).

611.   Bretschneider, E. **History of European Botanical Discoveries in China**. Leipzig, Germany: Zentral Antiquariat der Deutschen Demokratischen Republik, 1962. 2v.

This reference source, which is a reprint of the 1898 edition, surveys botanical exploration in Eastern Asia by European travelers and botanists. There are name and plant indexes.

612.   Clarkson, Rosetta E. **The Golden Age of Herbs and Herbalists**. New York: Dover, 1972. 352p. ill. $5.95pa. ISBN 048622869X.

Originally entitled *Green Enchantment: The Magic Spell of Gardens* (1940), this is a fascinating survey and standard reference for garden history and herbal medicine from the Middle Ages to the eighteenth century. It is beautifully illustrated and can be enjoyed on all levels.

613.   Coats, Alice M. **The Book of Flowers: Four Centuries of Flower Illustration**. New York: McGraw-Hill, 1973. 208p. ISBN 0070114803.

Limited to flowers as represented on paper or vellum, this book covers the period from 1485 to 1850. Coats introduces her beautiful book by discussing flower-books from the sixteenth through the nineteenth centuries. There are 126 examples of splendid illustrations, all of them from famous herbals, journals, and the like. Text is oriented toward the history of botanical illustration; a historical bibliography is included. [R: Choice, Mar. 1974, p. 70]

614.   Coats, Alice M. **Flowers and Their Histories**. New York: McGraw-Hill, 1971. 348p. plates. (col.). bibliog. ISBN 0070114765.

This is a systematic historical account of familiar garden flowers; roses, alpine, and greenhouse flowers are not included. The book is divided into three parts: border flowers, herbs, and some short biographies of famous gardeners. There is a bibliography and an index of English names differing from the Latin. This is a good companion to the *Book of Flowers* (entry 613) and *Flowers in History* (entry 617).

615.   Coats, Alice M. **The Plant Hunters: Being a History of the Horticultural Pioneers, Their Quests and Their Discoveries from the Renaissance to the Twentieth Century**. New York: McGraw-Hill, 1969. 400p. ISBN 0070114757.

This book is concerned with approximately 500 professional gardeners, who traveled abroad for the purpose of collecting hardy garden plants. Chapters are arranged geographically and then chronologically; there is an extensive bibliography.

616.   Coats, Alice M. **Treasury of Flowers**. New York: McGraw-Hill, 1975. 118p. $14.95. ISBN 007011482X.

A companion volume and parallel source to the *Book of Flowers* (entry 613), this volume is compiled from selections taken from octavo-size publications, botanical periodicals, and ordinary gardening books not included in the *Book of Flowers*. "The two volumes provide the most complete portrayal of historical floral illustration published to date." [R: Choice, Feb. 1976, p. 1559]

617.   Coats, Peter. **Flowers in History**. New York: Viking Press, 1970. 264p. ISBN 067055894X.

This beautiful book is heavily illustrated, and like *Flowers and Their Histories* (entry 614), written in a companionable, interesting style. Chapters are devoted to particular flowers, 15 in all with the rose included, and there is a chapter on herbs.

618.   Daniels, Gilbert S. **Artists from the Royal Botanic Gardens, Kew**. Pittsburgh, Pa.: Hunt Institute for Botanical Documentation, Carnegie-Mellon University, 1974. 73p. illus. $3.00pa. ISBN 0913196177.

Although this is a catalog of an exhibition held at the Hunt Institute, it is included here because of the importance of Kew Gardens, one of the leading botanical gardens and scientific collections of the world. This catalog represents Kew's huge art holdings by introducing 39 artists and their botanical illustrations collected by the garden. Information about each artist includes biographical material, a portrait of the artist, and a sample of work with bibliographic citation. The chronology covers the range of Kew history from the seventeenth century to contemporary artists. This catalog is useful for the uninitiated and for identification and information on selected botanical illustrations.

619.   Darlington, W. **Memorials of John Bartram and Humphrey Marshall, with Names of Their Contemporaries**. Introduction by J. Ewan. New York: Hafner, 1967. 585p.

This facsimile of the 1849 edition has been called the most frequently cited book today dealing with eighteenth-century botanical history in the American Colonies. Contents include discussions of the progress of botany in North America, biographical sketches of John Bartram and Humphrey Marshall, and their correspondence with other well-known botanists and collectors of their day.

620.   Desmond, Ray. **Dictionary of British and Irish Botanists and Horticulturists Including Plant Collectors and Botanical Artists**. London: Taylor & Francis, 1977. 764p. ISBN 0850660890.

There are over 10,000 entries for botanists, plant collectors, and botanical artists that include biographical information, references in the literature, commemorative plant names, and location of collections, manuscripts, drawings, and portraits. This excellent source of information on the development of botany in Britain is recommended for botanical and historical libraries. [R: Nature, Dec. 1, 1977, p. 457]

621.   **Development of Botany in Selected Regions of North America before 1900**. Edited with an introduction by Ronald L. Stuckey. New York: Arno, 1978. 258p. $25.50. ISBN 0405107226. (Biologists and Their World).

This is a collection of papers written around 1900 and reprinted from recognized journals on the development of botany in the United States, principally in the eastern region. Articles discuss botany in New York, the District of Columbia, the South, and St. Louis, from 1635 to 1858.

622. **Dictionary of American Biography**. Published under the auspices of the American Council of Learned Societies. New York: Scribner, 1928-37. 20v. and index; repr., New York: Scribner, 1946. 17v. $1,100/set (includes 7 supplements). ISBN 0684173239.

This is designed along the lines of the English *Dictionary of National Biography* (entry 623) to include biographical and bibliographical information for all notable deceased Americans. Treatment is scholarly.

623. **Dictionary of National Biography**. Edited by Sir Leslie Stephen and Sir Sidney Lee. London: Smith, Elder, 1908-9; repr., New York: Oxford University Press, 1938. 22v. **Supplements**. New York: Oxford University Press, 1912-81. 7v. $998.00/set. ISBN 0198651015.

This is the most important reference for English biography. Scope covers all noteworthy inhabitants of the British Isles and the Colonies, exclusive of living persons, to 1970. Signed articles and comprehensive bibliographies are included.

624. **Dictionary of Scientific Biography**. Charles Coulston Gillispie, ed.-in-chief. New York: Scribner, 1970-80. 8v. $750.00/set. ISBN 0684169622.

"Published under the sponsorship of the American Council of Learned Societies with the endorsement of the History of Science Society." Biographical information covers scientists from all periods of history, excluding living persons, in astronomy, biology, chemistry, earth sciences, mathematics, and physics. Scope is international. Biographical information, complete bibliographies of the scientist's work, and a comprehensive discussion of their scientific contribution is included. Criteria for selection include the requirement that the scholar's work be "sufficiently distinctive to make an identifiable difference to the profession or community of knowledge."

625. Dodge, Bertha Sanford. **Plants That Changed the World**. Illustrated by Henry B. Kane. Boston: Little, Brown & Co., 1959. 183p. bibliog.

This very readable book describes some of the plant products, and adventures, that have helped to make history. Ornamental plants are not included; rather, the focus is on valuable plants whose introduction may be lost in the mists of history. A bibliography records sources of information.

626. Dunthorne, Gordon. **Flower and Fruit Prints of the 18th and Early 19th Centuries: Their History, Makers and Uses, with a Catalogue Raisonne of the Works in Which They Are Found**. Washington, D.C.: n.p., 1938; repr., New York: Da Capo, 1970. 275p. (Da Capo Press Series in Graphic Art, Vol. 6).

This beautiful book is written by a print lover who evaluates the quality of the prints from the technical as well as the aesthetic viewpoint; the scientific approach is outside its scope. The book is divided into two parts: the first discusses history and description of prints, including illustrations; the second has a catalog of early prints with complete descriptions of each print, artist, engraver, publisher, and other information necessary for identification. Indexes are included. This source may be used for verification as well as for locating prints of particular plants and artists. Blunt (entry 609), Coats (entries 613 and 616), and Nissen (entry 53) may be used as companion works.

627.   Duval, Marguerite. **The King's Garden**. Translated by Annette Tomarken and Claudine Cowen. Charlottesville, Va.: University Press of Virginia, 1982. 214p. $17.95. ISBN 0813909163.

This lively history surveys French botanical exploration and discovery for more than three centuries. There is an appendix listing the botanical gardens and arboretums of France.

628.   Ewan, Joseph A. "The Botanic Garden and the Book." In **Hortus Botanicus**. Compiled by I. MacPhail. Lisle, Ill.: Morton Arboretum, 1972.

This is an introductory essay on the history and development of the botanic garden.

629.   **Fifty Years of Botany: Golden Jubilee Volume of the Botanical Society of America**. Edited by W. C. Steere. New York: McGraw-Hill, 1958. 638p.

Some historical articles are included in this survey: "Early History of the Botanical Society of America," "Highlights of Botanical Exploration in the New World," "Botanical History," and reports of progress in selected areas over a fifty-year span.

630.   Ford, C. E. "Botany Texts: A Survey of Their Development in American Higher Education, 1643-1906." **History of Education Quarterly** 4, no. 1 (1964): 59-70.

This article surveys botany texts in American higher education for the years indicated.

631.   Gascoigne, Robert Mortimer. **A Historical Catalogue of Scientific Periodicals, 1665-1900, with a Survey of Their Development**. New York: Garland, 1985. 206p. bibliog. index. $27.00. ISBN 0824087526. (Garland Reference Library of the Humanities, Vol. 583).

This is a companion volume to *A Historical Catalogue of Scientists* (entry 632).

632.   Gascoigne, Robert Mortimer. **A Historical Catalogue of Scientists and Scientific Books from the Earliest Times to the Close of the Nineteenth Century**. New York: Garland, 1984. 1200p. $150.00. ISBN 0824089596.

Over 13,000 scientists, including botanists, from all countries, from antiquity to 1900, are covered in this major reference work. Biographical information, references to sources for additional information, scientific books, plus author and topical indexes are contained in this historical survey of the scientific literature.

633.   Gibbs, R. Darnley. "History of Chemotaxonomy." In **Chemotaxonomy of Flowering Plants**. Montreal: McGill-Queen's University Press, 1974. 4v. $150.00/set. ISBN 0773500987.

A brief history from the viewpoint of a well-known phytochemist.

634.   Goodale, George Lincoln. "The Development of Botany since 1818." In **A Century of Science in America with Special Reference to the American Journal of Science 1818-1918**. New Haven, Conn.: Yale University Press, 1918. (Yale University Mrs. Hepsa Ely Silliman Memorial Lectures).

This book commemorates the one hundredth anniversary of the founding of the *American Journal of Science* by Benjamin Silliman in 1818. After the first chapter which gives an account of the *Journal*'s beginnings, the remaining chapters discuss branches of sciences (geology, paleontology, petrology, mineralogy, chemistry, physics, zoology, and botany) that have been prominently included in the *Journal* from its inception.

635.   Goode, George B. "The Beginnings of Natural History in America." "The Beginnings of American Science." In **Annual Report of the Smithsonian Institution, 1897. Report of the U.S. National Museum, Pt. II**. Washington, D.C.: Government Printing Office, 1901.

These Presidential addresses (1886-87), delivered by the author to the Biological Society of Washington, offer an interesting and scholarly summary. Dr. Goode, Assistant Secretary of the Smithsonian Institution, in charge of the United States Museum, was a revered scientist and historian of science in America.

636.   Green, J. R. **A History of Botany, 1860 to 1900: Being a Continuation of Sachs's "History of Botany, 1530-1860."** New York: Russell & Russell, c1909, 1967. 543p.

One of the classic histories of botany; it is a companion and continuation for Sachs (entry 663). See Weevers (entry 670) for botanical history to 1945.

637.   Greene, Edward Lee. **Landmarks of Botanical History**. Edited by Frank N. Egerton. A Publication of the Hunt Institute for Botanical Documentation, Carnegie-Mellon University. Stanford, Calif.: Stanford University Press, 1983. 2v. $100.00/set. ISBN 0804710759.

Part 1 was first published in 1909, but has long been out-of-print. The second part, taken from Greene's 1915 unpublished manuscript, was brought out in 1983 with a new introduction, a short biography of Greene, and an appraisal of his somewhat controversial contributions to botany.

The main body of the two volumes is concerned with Greene's identification and discussion of the landmarks of botany. Part 1 deals with an introduction to the philosophy of botanical history; part 2 discusses the Italian forefathers of the fifteenth century; and appendixes outline pre-Grecian, medieval, and seventeenth-century botany. These volumes will be of interest to historians, taxonomists, and botanists. [R: Science, Dec. 9, 1983, pp. 1113-14]

638.   Harshberger, John William. **The Botanists of Philadelphia and Their Work**. Philadelphia: Published by the author, 1899. 457p.

This contribution to the history of botany in America encompasses the area within a radius of sixty miles of Philadelphia and includes biographical sketches for most of the botanists who lived near Philadelphia. There are 40 illustrations including portraits of scientists and their living quarters; appendixes listing members of botanical clubs and societies; and historical accounts of scientific journals and landmarks.

639.   Harshberger, John William. **Phytogeographic Survey of North America: A Consideration of the Phytogeography of the North American Continent, Including Mexico, Central America and the West Indies, Together with the Evolution of North American Plant Distribution**. New York: G. E. Stechert, 1911; repr., Monticello, N.Y.: Lubrecht and Cramer, 1958. 790p. $35.00. ISBN 3768200035.

The contents include part 1, History and Literature of the Botanic Works and Explorations of the North American Continent; part 2, Geographic, Climatic and Floristic Survey; part 3, Geologic Evolution, Theoretic Considerations and Statistics of North American Plants; and part 4, North American Phytogeographic Regions,

Formations, Associations. A historical survey of the phytogeography of North America makes this an interesting and useful book for the botanist and the historian of science.

640.   Harvey, John Hooper. **Mediaeval Gardens**. Beaverton, Oreg.: Timber Press, 1982. 199p. ill. $34.95. ISBN 0917304691.
This elegant and scholarly book puts together a critical and detailed account of British gardens from 1000 to 1500 A.D. There is a wealth of illustrations, interesting commentary, and careful documentation; this book has wide appeal. [R: QRB, Mar. 1983, p. 93]

641.   Haughton, Claire S. **Green Immigrants: The Plants That Transformed America**. New York: Harcourt Brace Jovanovich, 1978. 464p. $5.95pa. ISBN 0156364921.
Haughton very effectively meets her aim of relating the "history and romance, the legend and folklore, of nearly one hundred growing plants, telling where they came from, how they arrived here, and what has happened to them since."

642.   Hawks, Ellison, and George S. Boulger. **Pioneers of Plant Study**. London: Sheldon Press, 1928; repr., New York: Arno, 1969. 288p. $17.00. ISBN 0836911393.
This is an episodic history of selected botanists from antiquity to the nineteenth century, their investigations, and important botanical institutions; 18 portraits are included.

643.   Heiser, Charles Bixler, Jr. **Seed to Civilization: The Story of Man's Food**. 2nd ed. San Francisco: W. H. Freeman & Co., 1981. 254p. $26.95; $12.95pa. ISBN 0716712644; 0716712652pa.
Although mainly concerned with the development of agriculture, this book is of general interest in its discussions of the history of man's food and particular food plants.

644.   Henrey, Blanche. **British Botanical and Horticultural Literature before 1800: Comprising a History and Bibliography of Botanical and Horticultural Books Printed in England, Scotland and Ireland from the Earliest Times until 1800**. New York: Oxford University Press, 1975. 3v. $195.00/set. ISBN 0686787390.
Volume 1 deals with the history and bibliography of the sixteenth and seventeenth centuries; volume 2 discusses eighteenth-century botanical history; and volume 3 is the bibliography for eighteenth-century books. This is a comprehensive, authoritative source that is important historically, as well as bibliographically, for the British Isles.

645.   Hill, A. W. "History and Functions of Botanic Gardens." **Annals of the Missouri Botanical Garden** 2 (1915): 185-240.
The title is descriptive of the contents, with the discussion focusing on the founding of various famous botanical gardens. Twelve plates of gardens, maps, and herbals are included.

646.   **Historiae Naturalis Classica**. Vol. 1- . Weinheim, Germany: J. Cramer; distr., Riverside, N.J.: Stechert, 1959- . irregular. price varies.
This is a series of facsimile reprints of important works in botany and zoology.

647. **History in the Service of Systematics: Papers from the Conference to Celebrate the Centenary of the British Museum (Natural History) 13-16 April, 1981**. Edited by Alwyne Wheeler and James H. Price. London: Society for the Bibliography of Natural History, 1981. 164p. $20.00. ISBN 0901843059. (Special Publication, No. 1).

One reviewer of this celebratory volume reported that this work illuminated the crucial role historical research plays in escaping the cul-de-sacs taxonomists often encounter and in particular, he called it a "delightful summary view of history in the service of systematics." [R: Huntia 5, no. 1, 1983, pp. 80-82]

648. Humphrey, Harry Baker. **Makers of North American Botany**. New York: Ronald Press, 1961. 265p. (Chronica Botanica, No. 21).

Short biographies (one to three pages) of 122 famous North American botanists are included; there are no portraits, although references to their obituaries are provided. The years up to 1958 are covered.

649. Hutchinson, J. "Short History of the Classification of Flowering Plants, before Linnaeus and After." In **The Families of Flowering Plants Arranged According to a New System Based on Their Probable Phylogeny**. 3rd ed. New York: Oxford University Press, 1973; repr., Forestburgh, N.Y.: Lubrecht and Cramer, 1979. ill. $63.00. ISBN 387429160X.

Provides a brief history of classification of flowering plants.

650. **Index Herbariorium. Pt. 2: Index of Collectors**. Utrecht, Netherlands: Bohn, Scheltema and Holkema, 1983. $34.25. ISBN 9031305855. (Regnum Vegetabile, Vols. 2, 9, 86, 93, and 109).

This is a list of plant hunters, issued irregularly as part of the Regnum Vegetabile series, with vital dates, collection speciality, location of vouchers, and sources.

651. Johnson, Dale E. "Literature on the History of Botany and Botanic Gardens 1730-1840: A Bibliography." **Huntia** 6, no. 1 (1985): 3-121.

The entire issue is devoted to a bibliography of literature on the study of plants. There are 426 entries, short titles, person and subject indexes. Information includes author, title, edition, imprint, notes, contents, and reviews. Biographies are excluded, but works on the history of botanic gardens are included. This is an extremely useful synthesis.

652. Kelly, Howard A. **Some American Medical Botanists Commemorated in Our Botanical Nomenclature**. Troy, N.Y.: Southworth Company, 1913; repr., New York: Dover, 1977. 215p. $25.00. ISBN 0893411450.

In this interesting contribution, Kelly provides brief sketches of the lives of 30 early medical men "who have been honored and immortalized by these floral tributes." Illustrations include portraits of botanists/physicians and the flowers named after them.

653. King, Ronald. **Botanical Illustration**. New York: Clarkson N. Potter, 1979. 16p. ill. (col.). $14.95; $6.95pa. ISBN 0517535254; 0517535262pa.

This work relies heavily on *The Art of Botanical Illustration* (entry 609) by Wilfrid Blunt, to whom King dedicated his book. The 40 full-page color plates are beautifully done. King, who includes two twentieth-century artists in his collection, suggests that

Blunt was too pessimistic when he predicted that "the great days of botanical art lie behind us."

654.   Lemmon, Kenneth. **The Golden Age of Plant Hunters**. Cranbury, N.J.: A. S. Barnes, 1969. 229p.
   This is another interesting book on botanical travelers and their place in history; enjoyable competition for Coats (entry 615) and Tyler-Whittle (entry 669).

655.   Lindau, Gustav, and Paul Sydow. For additional biographical information relevant to mycology, see entry 46.

656.   Meisel, Max. See entry 49 for an annotation for this indispensable guide to early American natural history.

657.   Miller, Amy Bess Williams. **Shaker Herbs: A History and a Compendium**. New York: Clarkson N. Potter, 1976. 272p. ISBN 0517524945.
   This well-done history of Shaker herbs is a worthy addition to botanical collections, public or academic. The book is divided into two parts: discussions of herbs from various Shaker communities and a compendium of herbs detailing their use. A glossary, bibliographies, and an index are included. [R: ARBA, 1978, entry 1298]

658.   Morton, A. G. **History of Botanical Science: An Account of the Development of Botany from Ancient Times to the Present Day**. New York: Academic, 1981. 474p. $55.00; $27.00pa. ISBN 0125083807; 0125083823pa.
   According to the preface, this "modern" history has been written "in a form sufficiently concise and uncluttered by detail to be comprehensible to readers who are not botanists and have no more than a general interest in the history of science whilst at the same time giving botanists an adequate account of the development of the principles and the factual basis of their science." This book is written with "great care and scholarship and it traces in detail the emergence of philosophical concepts within the science of botany." [R: Nature, Apr. 7, 1983, p. 554]

659.   Nissen, Claus. **Herbals of Five Centuries: 50 Original Leaves from German, French, Dutch, English, Italian and Swiss Herbals**. Zurich, Switzerland: L'Art Ancien, 1958. 86p. (50 plates and text).
   A brief discussion of medical history and bibliography includes 50 plates from original herbals.

660.   Reed, H. S. **A Short History of the Plant Sciences**. Waltham, Mass.: Chronica Botanica, 1942. 320p.
   This is another of the classic histories of botany. Chapters include discussions on gardeners and herbalists, the chronology of botanical work, and various topics of importance: morphology, cytology, fixation, metabolism of nitrogen, etc. There is a short list of significant discoverers, classifiers, specialists, and exponents of botany for beginners.

661.   Rogers, A. D., III. **American Botany, 1873-1892: Decades of Transition**. Princeton, N.J.: Princeton University Press, 1944. 340p.

This book covers a crucial period of transition and development in the course of American botany. This very interesting account discusses prominent botanists, their explorations, controversies, and research, relying heavily on primary sources.

662. Rogers, Donald Philip. **A Brief History of Mycology in North America**. Amherst, Mass.: Second International Mycological Congress, Inc., 1977; repr., Tampa, Fla.: Mycological Society of America, 1981. 86p.

Reprinted in "augmented form," this history presents mycology from the American viewpoint, with thumbnail sketches of prominent early mycologists and their most important research. For a more comprehensive history of the topic with a British/Continental flavor, see Ainsworth's *Introduction to the History of Mycology* (entry 598).

663. Sachs, Julius von. **History of Botany (1530-1860)**. rev. ed. Edited by I. B. Balfour. Translated by H. E. Garnsey. München, Germany: R. Oldenbourg, 1875; repr., New York: Russell & Russell, 1967. 568p. $11.00. ISBN 084621797X.

This is one of the classic histories of the botanical sciences; it is continued by Green (entry 636), and then by Weevers (entry 670).

664. **A Short History of Botany in the United States**. Edited by Joseph A. Ewan. New York: Hafner, 1969. 174p. $8.95. ISBN 0028443608.

This is a historical review of the centers of botanical activity in the United States into the mid-twentieth century.

665. Sitwell, Sacheverell, and Wilfrid Blunt. **Great Flower Books, 1700-1900: A Bibliographical Record of Two Centuries of Finely Illustrated Flower Books**. Bibliography edited by Patrick M. Synge. London: Collins, 1956. 94p. plus index. 36 plates.

The major purpose is to describe some of the great flower books of the eighteenth and nineteenth centuries.

666. Stearn, William T. "Historical Introduction." In **International Code of Nomenclature for Cultivated Plants**. London: Royal Horticultural Society, 1953.

This is an outline of the early history of the codes of botanical nomenclature.

667. Steenis-Kruseman, M. J. van. "Malaysian Plant Collectors and Collections; Being a Cyclopaedia of Botanical Exploration in Malaysia and a Guide to the Concerned Literature Up to the Year 1950." In **Flora Malesiana**, Ser. 1, Vol. 1. Djakarta, Indonesia: Noordhoff-Kolff, 1950. 639p. ill. index.

Data include collectors' biographical information, portraits, itinerary, location of collections, and literature. This compilation provides hard-to-find information for botanical collectors in Malaysia and is a good candidate for botanical research libraries.

668. "25 Years of Botany, 1947-1972." **Annuals of the Missouri Botanical Garden** 61, no. 1 (1974): 1-261.

This review of the progress in various fields of botany following World War II is written by 14 prominent contributors.

669. Tyler-Whittle, Michael Sydney. **The Plant Hunters: Being an Examination of Collecting with an Account of the Careers and the Methods of a Number of Those Who Have Searched the World for Wild Plants**. Philadelphia: Chilton, 1970. 281p.

This book discusses "why, how and where some plants have been collected, with an account of a few of the better-known collectors." This is an amusing and interesting contribution appropriate for scientific and public libraries.

670.   Weevers, Theodorus. **Fifty Years of Plant Physiology**. Utrecht, Netherlands: Bonn, Scheltema and Holkema, 1949. 308p.

This is a continuation of the history of botany begun by Sachs (entry 663) and Green (entry 636). Weevers discusses botanical history from 1895 to 1945 with attention to European, principally Dutch, botanical literature of the period covered.

9 Textbooks

Following are selected examples of especially important textbooks. Texts are usually defined as tertiary scientific literature, meant for instruction and to provide information to aid in the understanding of an area. Textbooks can be written appropriately for any level from the elementary, which assumes no knowledge of the topic, to the advanced, which may take on the attributes of treatises due to their thorough coverage, development, and treatment of the subject. Although there will be some overlap, in general, treatises are annotated with reviews in chapter 3; other textbooks are discussed wherever they are relevant to chapter content. For the most part, textbooks included in this section are not annotated—their role and importance in supporting and supplying instruction and definitive information may be taken for granted.

671. **Anatomy of the Dicotyledons.** 2nd ed. Edited by C. R. Metcalfe and L. Chalk. New York: Oxford University Press, 1980-83. 2v. Vol. 1: $68.00. ISBN 0198543832. Vol. 2: $75.00. ISBN 0198545592.

672. Benson, Lyman. **Plant Classification**. 2nd ed. Lexington, Mass.: Heath, 1979. 901p. $31.95. ISBN 0669014893.

673. Bernier, Georges, Jean-Marie Kinet, and Roy M. Sachs. **The Physiology of Flowering**. Boca Raton, Fla.: CRC Press, 1981-85. 3v. Vol. 1: $68.00. ISBN 0849357098. Vol. 2: $84.00. ISBN 0849357103. Vol. 3: $99.00. ISBN 084935711X.

For complete encyclopedic data on control and regulation of flowering see entry 316.

674. **Biochemistry of Plants; A Comprehensive Treatise**. Edited by P. K. Stumpf and E. E. Conn. New York: Academic Press, 1980-81. 8v. $551.50/set.

675. Bold, Harold C., et al. **Morphology of Plants and Fungi**. 4th ed. New York: Harper & Row, 1980. 819p. $35.95. ISBN 0060408480.

676. Dahlgren, Rolf M. T., and H. Trevor Clifford. **The Monocotyledons: A Comparative Study**. New York: Academic, 1982. 378p. $98.50. ISBN 0122006801. (Botanical Systematics, Vol. 2).

677. Dahlgren, Rolf M. T., H. Trevor Clifford, and Peter F. Yeo. **The Families of the Monocotyledons: Structure, Evolution, and Taxonomy**. New York: Springer-Verlag, 1985. 520p. $98.00. ISBN 038713655X.

678. Daubenmire, Rexford F. **Plant Geography: With Special Reference to North America**. New York: Academic, 1978. 338p. $47.00. ISBN 012204150X.

679. Esau, Katherine. **Anatomy of Seed Plants**. 2nd ed. New York: John Wiley, 1977. 550p. $39.45. ISBN 0471245208.

680. Foyer, Christine H. **Photosynthesis: Cell Biology**. New York: John Wiley, 1984. 219p. $29.95. ISBN 0471864730. (Cell Biology, Vol. 1).

681. **The Fungi: An Advanced Treatise**. Edited by Geoffrey C. Ainsworth, Frederick K. Sparrow, and Alfred S. Sussman. New York: Academic, 1965-73. 4v. $358.00/set. ISBN 0685051285.

682. Goodwin, Trevor Walworth, and Eric Ian Mercer. **Introduction to Plant Biochemistry**. 2nd ed. New York: Pergamon Press, 1982. 400p. ill. $99.00. ISBN 0080249221.

683. Grant, Verne. **Genetics of Flowering Plants**. New York: Columbia University Press, 1975. 514p. ill. $52.50; $22.00pa. ISBN 0231036949; 0231083637pa.

684. Grant, Verne. **Plant Speciation**. 2nd ed. New York: Columbia University Press, 1981. 544p. $52.50; $18.50pa. ISBN 0231051123; 0231044607pa.

685. Gunning, Brian E. S., and Martin W. Steer. **Ultrastructure and the Biology of Plant Cells**. London: Edward Arnold, 1975. 320p. ISBN 0713124946.
    An excellent collection of electron micrographs.

686. Hall, Timothy C., and Jeffrey W. Davies. **Nucleic Acids in Plants**. Boca Raton, Fla.: CRC Press, 1980. 2v. Vol. 1: $76.00. ISBN 0849352916. Vol. 2: $71.00. ISBN 0849352924.

687. Ingold, Cecil Terence. **The Biology of Fungi**. 5th ed. London: Hutchinson, 1984. 176p. ISBN 0091545919.

688. Krebs, Charles J. **Ecology: The Experimental Analysis of Distribution and Abundance**. 3rd ed. New York: Harper & Row, 1985. 704p. $33.50. ISBN 0060437782.

689. **Photosynthesis**. Edited by Govindjee. New York: Academic, 1982. 2v. Vol. 1: $59.00. ISBN 0122943015. Vol. 2: $79.00. ISBN 0122943023.

690. **Plant Biochemistry**. 3rd ed. Edited by James Bonner and Joseph E. Varner. New York: Academic, 1976. 925p. $42.50. ISBN 0121148602.

691. **Plant Physiology: A Treatise**. Vol. 1- . New York: Academic, 1960- . irregular. (Vol. 8: 1983. $70.00. ISBN 0126686084). Also available by subscription.

692. **The Plant Viruses**. Vol. 1- . Edited by R. I. B. Francki. New York: Plenum, 1985- . (Vol. 1: 297p. $49.50. ISBN 0306419580). Subseries within *The Viruses*, edited by H. Fraenkel-Conrat.

693. Raven, Peter H., et al. **Biology of Plants**. 3rd ed. New York: Worth, 1981. 685p. $36.95. ISBN 0879011327.

694. Ricklefs, Robert E. **Ecology**. 2nd ed. New York: Chiron, 1979. 966p. $29.95. ISBN 0913462071.

695. Salisbury, Frank B., and Cleon W. Ross. **Plant Physiology**. 3rd ed. Belmont, Calif.: Wadsworth, 1985. 540p. ISBN 0534044824.

696. Schultes, Richard E., and Albert Hofmann. **The Botany and Chemistry of Hallucinogens**. 2nd ed. Springfield, Ill.: Charles C. Thomas, 1980. 464p. $46.50. ISBN 0398038635.

697. Walter, Heinrich. **Vegetation of the Earth and Ecological Systems of the Geobiosphere**. 3rd ed. New York: Springer-Verlag, 1985. 340p. $17.00pa. ISBN 0387137483.

698. Weir, T. Elliot, et al. **Botany: An Introduction to Plant Biology**. New York: John Wiley, 1982. 720p. $37.45. ISBN 047101561X.

10 Key Publishers, Services, and Important Series

As in every field, work in the botanical sciences is reported by a wide selection of publishers and producers who issue materials and services in various formats and at various frequencies. Certain publishers specialize in botany; it is useful for botanists of any calling, at any level, to be aware of the groups that can be expected to publish botanical materials of importance. Following are selected examples of especially significant publishers, services, and series crucial to the study of botany. Societies and associations are not included here; see chapter 6 for lists of professional organizations and their publications of consequence.

699. **Academic Press, Orlando, Fla.**
This respected publisher is well-known for scholarly books on all subjects. Relevant to botany are the following series:

Experimental Botany: An International Series of Monographs. Recent contributions to this series include:

The Experimental Biology of Bryophytes. 1984. $71.50. ISBN 0122263707.

Plant Histochemistry. 1984. $46.00. ISBN 0122732707.

Systematics Association Special Volumes series. One of the newer issues in this series is *Databases in Systematics*. 1984. $54.50. ISBN 0120530406.

700. **BioSciences Information Service (BIOSIS), Philadelphia**.

BIOSIS has been serving the information needs of the world community of bio-scientists since 1926 when it began *Biological Abstracts*. BIOSIS is an independent, not-for-profit organization whose mission is to provide secondary sources for access of primary information on an international basis to scientists, educational and research institutions, government agencies, and industrial corporations. According to a report from BIOSIS, in 1985, coverage will total 440,000 items originally published in more than 9,000 serials and other publications worldwide. The entire BIOSIS file now contains over six million reports making it the world's largest collection of abstracts and citations for biology in the English language. Beginning in 1985, University Microfilms International is the official source for document delivery for materials abstracted in the BIOSIS database. BIOSIS offers printed reference publications like *Biological Abstracts* and *Biological Abstracts/RRM*; speciality publications like *Abstracts of Mycology*; computer services; magnetic tape services; educational programs and materials; and microform services. For more information on the BIOSIS enterprise, see *Biological Abstracts: BIOSIS—The First Fifty Years*, edited by William C. Steere (New York: Plenum, 1976, 250p., ill., $59.50, ISBN 0306309157).

701. **Bohn, Scheltema and Holkema, Utrecht, The Netherlands**.

This well-known publisher issues Regnum Vegetabile, edited by Frans A. Stafleu, a series of well over 100 publications for the use of plant taxonomists, published under the auspices of the International Association for Plant Taxonomy. Examples of some of these important series and books are *Index Herbariorum*, *Index Muscorum*, *Index to Plant Chromosome Numbers*, *International Code of Botanical Nomenclature*, and *Taxonomic Literature*.

702. **British Museum (Natural History), London**.

Examples of their publications include:

The Bulletin of the British Museum (Natural History) is issued in five series: Zoology, Entomology, Botany, Geology, and Historical. Parts are issued irregularly, each complete in itself, individually priced. The Botany series (ISSN 0068-2292) began in 1949.

Flora of Iraq. Vol. 1- . 1966- .

Flora Zambesiaca. Vol. 1- . 1960- .

List of British Vascular Plants. 1958; repr., 1982.

703. **CAB International, formerly Commonwealth Agricultural Bureaux (CAB), U.K.**

Founded in 1929, CAB is one of the leading world services for agricultural science, providing: (1) information service, covering every branch of agricultural science and related aspects of applied biology, sociology, and economics; (2) identification service for insects, fungal and bacterial diseases of plants, helminth pests of animals and man, and plant-parasitic nematodes; and (3) biocontrol service to manage animal and plant pests using biotic agents. CAB publishes research journals, abstracting periodicals (available in print or online as CAB Abstracts Database), handbooks, and guides of interest to the worker in applied and basic botanical science areas. One of the CAB affiliates is the CAB International Mycological Institute, which houses a unique culture collection and library. Other activities include culture collection and industrial services, online services, document delivery, training courses, consultancy and contract services, and a respected series of publications of books and journals in the field of mycology.

704. **Cambridge Scientific Abstracts, Bethesda, Md.**
This publisher issues a wide range of abstracting journals and online databases. Two of interest to botanists are *Microbiology Abstracts* (entry 91) and *Virology Abstracts* (entry 96) annotated in chapter 2.

705. **Cambridge University Press, New York and Cambridge, England.**
One of the oldest university presses in the world, Cambridge University Press has a well-deserved highly respected reputation as a publisher of scholarly books in all areas of the arts and sciences. An example of the series issued by this press is Symposia of the British Mycological Society; the 1983 symposium, *Ecology and Physiology of the Fungal Mycelium*, was published in 1984.

706. **Elsevier Scientific, New York.**
One of the prestigious series of interest to botanists from this publisher is Ecosystems of the World (David W. Goodall, ed.-in-chief). *Tropical Rain Forest Ecosystems*, edited by Frank B. Golley (1983, 329p., $113.00, ISBN 0444419861), is volume 14A of this series.

707. **F. Flüch-Wirth, CH-9053 Teufen, Switzerland.**
"Krypto News," published by this respected company, an international bookseller for botany and natural sciences, provides bibliographical documentation of recently published or forthcoming botanical literature of interest to botanical institutions and libraries. Special botanical catalogs are distributed from time to time on research collections, rare materials, taxonomic literature, and the like. The vendor also offers a search service for out-of-print and antiquarian books.

708. **Her Majesty's Stationery Office, London.**
Some of the very important publications emanating from this office are:

Kew Bulletin. Vol. 1- . 1887- . quarterly. price varies. Most of these original articles deal with vascular plants and mycological systematics. They are written by staff of the Royal Botanic Gardens at Kew, an internationally famous botanical garden and one of the oldest.

Other series are the *Kew Record of Taxonomic Literature Relating to Vascular Plants, Kew Bulletin Additional Series*, Kew Bibliographies, and *Notes from the Royal Botanic Garden*, Edinburgh.

709. **Houghton Mifflin, Boston.**
The Petersen Field Guide series, named after the famous naturalist Roger Tory Peterson, is one of this publisher's series of interest to botanists. These guides are beautifully designed for field use and written by experts. For some examples, see entry 534.

710.   **Hunt Institute for Botanical Documentation, Carnegie-Mellon University, Pittsburgh, Pa.**

This prestigious Institute and library has a long history of scholarly research and publishing activities in the botanical sciences. Some of their most important publications are *Botanico-Periodicum-Huntianum*, *Catalogue of Botanical Books in the Collection of Rachel McMasters Hunt*, *Huntia: A Journal of Botanical History*, and *Bulletin of the Hunt Institute for Botanical Documentation*. Their current projects include exhibitions of botanical books and paintings and various taxonomic and reference publications.

711.   **Institute for Scientific Information (ISI), Philadelphia.**

ISI is one of the larger multidisciplinary indexing services offering various printed and online databases for the sciences. Some of the indexes produced by ISI, discussed in chapter 2, are *Index to Scientific Reviews* (entry 88), *Current Contents* (entries 78 and 79), and *Science Citation Index* (entry 95).

712.   **Junk Publications, The Hague, Netherlands.**

Distributed by Kluwer-Academic, Boston, two series of import are *Handbook of Vegetation Science* (entry 244) and *Tasks for Vegetation Sciences* (entry 367).

713.   **Knopf Inc., New York.**

The Audubon Field Guides emanate from this publisher. These guides are comprehensive, easy to use in the field by the novice or expert, and written under the auspices of the Audubon Society. See entry 459.

714.   **Meckler Publishing, Westport, Conn. and London.**

This publisher offers a massive microfiche collection of important references to taxonomic botany; for an example, see *Taxonomic Literature* (entry 117). Microfiche copies of significant collections in herbaria plant type collections follow:

California Academy of Sciences Vascular Plant Type Collection.

New York Botanical Garden Vascular Plant Type Collection.

Types and Special Collections of the Herbarium of the Academy of Natural Sciences of Philadelphia.

United States National Arboretum Vascular Plant Type and Cultivar Collection.

United States National Herbarium, Smithsonian Institution Vascular Plant Types.

Other publications include journals and microfiche publications of important collections in the British Museum (Natural History), U.S. Department of Agriculture, etc.

715.   **The New York Botanical Garden, Bronx, N.Y.**

One of the most prestigious botanical research institutions and botanical libraries in the world, the New York Botanical Garden puts out several highly respected publications including:

Addisonia. Vols. 1-24. 1916-64. Back issues $10.00/vol. Established with the specific purpose of illustrating by color plates the plants of the United States and plants flowering at The New York Botanical Garden.

Advances in Economic Botany (entry 229), *Botanical Review* (entry 150), *Bulletin of the Torrey Botanical Club* (entry 153), *Brittonia* (entry 151), and *Economic Botany* (entry 156) are discussed in chapter 3.

Contributions from the New York Botanical Garden. 1899-1933. Vol. 15- . 1985- . irregular. price varies. ISSN 0736-0509.

Flora Neotropica. Vol. 1- . 1964- . irregular. $26.00/yr. ISSN 0071-5794.

Intermountain Flora. Vol. 1- . 1972- . irregular.

Memoirs of the New York Botanical Garden. Vol. 1- . 1900- . irregular. $16.00/yr. ISSN 0077-8931.

North American Flora. Vol. 1- . 1905- . irregular. $25.00/yr. ISSN 0078-1312.

716. Oxford University Press, Fair Lawn, N.J.

Oxford University Press publishes a wide variety of scholarly books in all areas of the arts, humanities, and sciences. One of its important series is the premier symposia in plant biochemistry, the Annual Proceedings of the Phytochemical Society of Europe, now in its 26th volume, *Plant Products and the New Technology* (1985, ISBN 0198541805). This series has achieved an international reputation as a definitive reference on a wide variety of topics on all aspects of plant biochemistry.

717. Pergamon Press, New York.

This international publisher issues research and review journals, major reference works, textbooks and monographs covering every field of academic inquiry; political publications; specialist computer software; and books for professionals. Of interest to botanists is the database, available in print or online, CURRENT AWARENESS IN BIOLOGICAL SCIENCES (CABS), storing current information taken from over 3,000 primary international journals. CURRENT ADVANCES IN PLANT SCIENCE (entry 77) is part of this database.

718. Smithsonian Institution Press, Washington, D.C.

Smithsonian Contributions to Botany. No. 1- . 1969- . irregular. ISSN 0081-024X. (No. 56, 1984). This series reports basic research, giving accounts of new discoveries in small papers or full-scale monographs.

719. Springer-Verlag, Berlin, Heidelberg, New York, Tokyo.

This publisher is highly visible in the botanical sciences. Two of their most important publications are the following series:

Ecological Studies. Vol. 1- . 1973- . irregular. ISSN 0070-8356. This monographic series deals with all aspects of ecology. Volume 49, *Forest Ecosystems in Industrial Regions*, was published in 1984.

Encyclopedia of Plant Physiology is annotated in entry 241.

720.   **University of California Press, Berkeley, Calif.**

University of California Publications in Botany. Vol. 1- . 1902- . irregular. price varies. ISSN 0068-6395. This is a long-standing series of individually priced monographs dealing with a single subject, covering all aspects of botany.

Index

Reference is to entry number. The letter "n" is used to designate citations to items found in annotations. The letter "p" is used to designate references to material in text.

Wiley, Edward O., 596
Wilkinson, Robert E., 505
Williams, Martha E., 379
Willis, John Christopher, 284n, 310, 549n
WILSONLINE, 211n
Winter Botany, 550
Winters, Wendell D., 10
Wooten, Jean W., 493
Wordsell, W. C., 104
*World Directory of Collections of
    Cultures of Microorganisms*,
    393
*World Guide to Scientific Associations
    and Learned Societies*, 399
World List, 210
*World List of Scientific Periodicals Pub-
    lished in the Years 1900-1960*, 210
World of Learning 1986, 400
World Patents Abstracts, 223n
World Pollen Flora, 553
Wormersley, J. S., 371
WPI Gazette Service, 223n

Wright, Nancy D., 9
Wyatt, H. V., 19
Wyllie, Thomas D., 248
Wynar, Bohdan S., 2

Yarrow, D., 461
Yearbooks, 231-32, 234, 239, 250, 257
Yeast: Characteristics and Identification,
    461
Yeasts, 461n
Yeasts—A Taxonomic Study, 554
Yeo, Peter F., 677

Zanoni, Thomas, 66
Zeitschrift für Pflanzenkrankheiten, p. 3
Zeitschrift für Pflanzenphysiologie, p. 4,
    166n
Zohary, Michael, 526n, 551n, 555
Zweifel, Frances W., 372
Zweig, Gunter, 330